Physics in the Arts

P.U.P.A. Gilbert and Willy Haeberli

University of Wisconsin - Madison

Table of Contents

Introduction
LIGHT AND SOUND

Through light and sound we perceive the world. We have perfected several ways of communicating with each other, and these all use either light or sound, or both. The highest form of communication is art, and all of the arts use light and/or sound as their expression medium. The visual arts use light, so we can see the works of art, music uses sound so we can hear it, ballet, movies, music videos with computer graphics use both.

A deep yet accessible analysis of the physics of light and sound, and how our eyes and ears detect them, is not only intellectually enjoyable, but also useful to understand and interpret the world in which we live, all the phenomena that take place around us, and how we perceive them. In short, how we interface with our planet, its inhabitants, and their creations. Understanding the physics of light and sound may also increase the appreciation for works of art and for art itself, and even stimulate the artists among the readers to deepen their knowledge of their media, of how people interface with them, and perhaps improve their art production.

Deriving its name from *phusis*, the Greek for "nature", physics is the science that studies naturally occurring phenomena concerning energy or matter. If we can understand a natural phenomenon, then we have contributed to mankind's general knowledge, but often we can also harness it and use it to develop a better life for ourselves, and all others on this planet. We think this exciting concept can be exported to the arts, and hopefully improve art productions through the understanding of the physical phenomena underlying them. This very thought motivates our desire to teach physics to artists and humanists and to communicate the physics concepts in simple terms to a more general audience.

Both sound and light are based on waves. These are different kinds of waves, mechanical waves in sound and electromagnetic waves in light, but they both oscillate as the surface of a water pond, after a stone has been thrown in the pond. Newton first coined the word *spectrum* of visible light, and then subdivided it into seven colors. Why seven? This choice was completely arbitrary, but he chose seven by analogy with the seven notes within an octave of the musical scale. We now know that the normal human eye can distinguish more than 10 million colors! For simplicity, however, we often divide the visible spectrum into three colors only: red, green and blue.

Have you ever wondered why the human eye is sensitive to the specific radiation called "light"? Although the sun emits radiation, which, after filtering through the atmosphere, comprises the visible range, there is also an awful lot of infrared and ultraviolet that reach the earth surface. Why don't we see in those ranges as well? In the visible range radiation has the most effective energy to interact with the objects of the world. It therefore best informs us on the structure and behavior of objects around us, without damaging these objects. At higher energy, in the ultraviolet and higher energies, radiation extracts electrons from atoms, therefore damaging molecules in the illuminated objects. At lower energies, in the infrared, radiation interacts so weakly with matter that it only provokes small vibrations that we perceive as heat, if at all. The measurable electromagnetic spectrum extends from high-energy gamma rays to low frequency radio waves. This covers more than 17 orders of magnitude in frequency (or energy, or wavelength). From the beginning to the end of what we call visible range (violet to red) the frequency varies by a mere factor of 2. In music, the frequency doubles in one octave. In the measurable electromagnetic spectrum the frequency doubles 56 times ($2^{56} = 10^{17}$). Poetically put, we can only "see one octave" on an idealized electromagnetic piano keyboard of 56 octaves! Yet so much happens in this narrow region because radiation of these energies strongly interacting with electrons generates non-flat, interesting absorption spectra, which in turn generate the millions of colors to which we are sensitive.

Compared to light, we hear as sound a much larger region of the usable frequencies. The frequency range for audible sound is 20-20,000 oscillations per second or 20-20,000 Hz. Higher frequencies, or ultrasounds, are used for medical imaging (10 MHz) and for communication by dolphins (up to 170 kHz), whales (up to 200 kHz) and bats (up to 120 kHz). Lower frequencies, or infrasounds, are produced by earthquakes, avalanches, volcanoes, aurora borealis and australis, nuclear tests, but also by elephants (14 Hz) for communication with other elephants within a 10 km distance.

The color and sound we perceive do not only depend on the physical, measurable stimuli, but also on the physiological and psychological response of our eyes, ears and brain to the stimuli. Color and sound, therefore, and are best described by psycho-physical parameters. These are, as we will describe in detail later, hue, saturation and brightness for color; pitch, loudness and timbre for sound. A trombone and a viola can immediately be distinguished by the listener even if they play the exact same tone. The attribute distinguishing them is called the timbre or, by anther analogy with light, "tone color".

Chapter 1
LIGHT AND LIGHT WAVES

We all have experience with light and its colors, variations in intensity, reflection, refraction, focusing, and light's characteristic of traveling along a straight line in a homogenous medium. We have, however, no intuitive feeling for the physical nature of light, what it is made of, and how it works. In this chapter we will discuss the dual nature of light—how it behaves as both waves and particles, and we will also discuss the electromagnetic spectrum, the speed of light in different media, and polarization.

Until the 1600s everyone thought light was particles. Then, in the 1600s competing theories of light were proposed by Christiaan Huygens who maintained that light was waves and Isaac Newton, who insisted it was particles. The debate continued, until finally in the 1900s it was agreed that the particles composing light, the photons, sometimes behave more as particles, other times more as waves. They, however, are always both particles and waves at the same time. This is termed the *dual nature of light*. Several phenomena can be

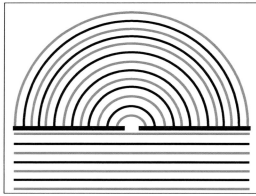

Figure 1.1. Schematic representation of the waves on water surface, seen from above. Red lines indicate crests, black lines troughs. A plane wave is propagating up from the bottom of the schematic. When it reaches a wall with an aperture, it propagates as a circular wave, centered at the aperture.

explained considering light as waves, while others are better explained considering light as particles[1].

More about waves. If you've ever been standing in shallow sea water, you probably noticed that you could not stop waves with your body. Waves just go around you. This phenomenon is called diffraction, and it lets waves go around obstacles almost undisturbed. Waves also go through a small opening and seem to diverge from there, as shown in Figure 1.1. Also, when the wakes from two boats combine standing waves may form. On the water surface, where both wake waves have a crest a stationary double-height crest results, while where both have troughs there is a stationary doubly deep trough. In the places, however, where one wake wave has a crest and the other has a trough, the water surface is flat, and

---

[1] Light behaves as waves in all experiments involving diffraction, interference, reflection and refraction. It behave as particles in Einstein's photoelectric effect, Planck's thermal radiation and Compton's scattering.

no wave is present. This phenomenon is called *interference. Constructive interference* generates doubly intense crests or troughs, *destructive interference* cancels a trough with a crest. If you now consider two slits, producing two sets of circular waves, you will see a similar interference effect. In Figure 1.2, where two black (troughs) or two red (crests) wave lines intersect there is constructive interference (solid dots), where one red and one black line intersect the interference is destructive (open dots).

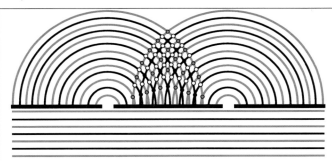

Figure 1.2. Schematic representation of the same waves going through two apertures and interfering. Where two red lines cross a crest of double height is formed, while at black line intersections troughs of double depths are formed by constructive interference (solid dots). Where a black line crosses a red line destructive interference makes the waves cancel each other, and the water surface is flat (open dots).

In a famous experiment done by Thomas Young in 1803, a very similar interference phenomenon takes place, using light instead of waves on the surface of water, and two light beams. Young's experiments proved for the first time that light is composed of waves. In his experiment, sunlight was reflected by a mirror, then filtered through a red filter, transmitted through a pinhole, and projected into a darkened room. At this point the red light beam was split into two light beams by a card, a thirtieth of an inch in thickness, held into the beam edgewise as shown in Figure 1.3. The two beams interfered constructively and destructively, as in Figure 1.2, and generated the pattern shown in Figure 1.3.

Figure 1.3. Young's experiment, demonstrating the wave nature of light. Left: a beam of sun light, filtered through a red filter (not shown) is made to pass through a pinhole. Another playing card is held edgewise, so the red beam is split into two beams, which interfere after the card to create the pattern on the right.

Often this experiment is thought to have been done by Young using two slits. That version of the same experiment, however, was not done until much later. The experiment of Figure 1.3 can easily be reproduced by the reader using a laser pointer as illuminating source, or sunlight, if the sun is shining and a darkened room is available. It is important to use a colored filter in the latter case: light of only one color (wavelength) must be used.

There is no conceivable way that would make particles accumulate more in certain spots and less in others, to generate the interference pattern of Figure 1.3. Young's experiment, therefore, proved for the first time that light indeed behaved as waves.

In this and all other experiments in which light is better interpreted as waves, there is a similarity between light and the crests and troughs of waves on the surface of water or waves traveling along strings. In light, what oscillates are the *electric and magnetic fields*. They both oscillate in the plane perpendicular to the travel direction. We all have some familiarity with these fields: the magnetic field is what makes a magnet attract a piece of iron and the electric field is responsible for the attraction of small pieces of paper to a comb that has been rubbed on wool (or dry hair).

As these fields oscillate in light waves, there is an alternation of places with a strong upward electric field, a strong downward field, and an electric field of zero. The pattern is similar to the one we see for waves on water: light moves with a velocity of propagation v and, as all other waves, it has an amplitude, a frequency, and a wavelength. Figure 1.4 graphically shows these quantities in a wave.

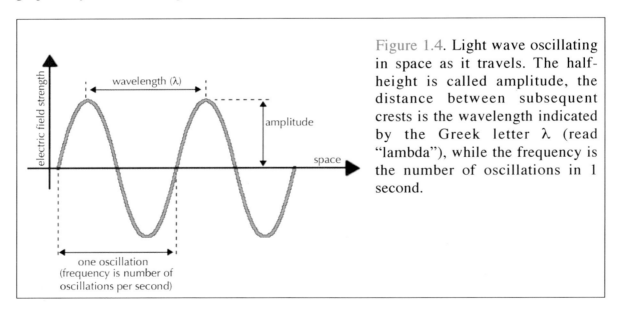

Figure 1.4. Light wave oscillating in space as it travels. The half-height is called amplitude, the distance between subsequent crests is the wavelength indicated by the Greek letter λ (read "lambda"), while the frequency is the number of oscillations in 1 second.

The wavelength λ varies for different colors. For example blue light has a shorter wavelength than red light, as shown in Figure 1.5.

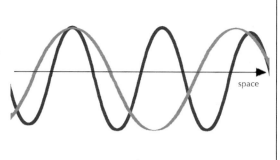

Figure 1.5. Two light waves oscillating as they travel. The amplitude is the same for both waves, while the wavelength for blue light is shorter than for red light.

SPEED OF LIGHT

The velocity of propagation of light depends on the medium through which it travels. *In vacuum*, and to a good approximation in air, the speed of light is 300,000,000 m/sec (or 3×10^8 m/s). This is a very considerable speed: by comparison, a fast car travels at 30 m/s, sound travels at 300 m/s, and a rocket leaving earth travels at only 10,000 m/s. The speed of light *in vacuum* is usually represented by the letter c (from the Latin word *celeritas*), so we write $c = 3 \times 10^8 \, m/s$.

In other transparent media, such as water, glass, and plexiglass, the speed (represented as v) of light is smaller than its speed in vacuum c. The actual value of v depends on the medium. The velocity v of light in a medium depends on *the index of refraction n* of the medium and can be found using the following equation:

$$v = \frac{c}{n}$$

The denser the medium, the larger its index of refraction and the slower light travels through it. For example: the index of refraction of plexiglass is $n_{plexiglass} = 1.5$ so the velocity of light in plexiglass is:

$$v = \frac{c}{n} = \frac{3x10^8}{1.5} = 2x10^8 m/s$$

The index of refraction of vacuum is $n_{vacuum} = 1$, the smallest index of refraction that exists. Therefore, light travels fastest in vacuum. Here is an interesting effect: imagine a transparent object with n > 1 surrounded by vacuum. When light traveling in vacuum traverses an object made of a transparent medium it slows down, then instantaneously it accelerates again when exiting the medium, without changing its energy.

ELECTROMAGNETIC SPECTRUM

Electromagnetic waves can have a large range of frequencies or of wavelengths; what we call light is the narrow range of wavelengths that can be detected by our eye, *400 nm to 700 nm* (*1 nm = one nanometer = 1×10^{-9} m*). This range of wavelengths is called *visible range*.

Electromagnetic waves with large wavelengths ($\lambda = 300$ m) are called radio waves and are received by radios. Waves with very short wavelengths are called ultraviolet light, or if shorter still are called x-rays and gamma-rays. Figure 1.6 shows energy, frequency, wavelength, and common name for the entire electromagnetic spectrum.

Figure 1.6.

The table below shows the wavelengths associated with the 7 colors commonly associated with the visible range (VIS). These are the colors originally identified by Newton, as dispersed by a prism. There are millions of colors in the spectrum, and not only 7, but Newton, by analogy with the 7 notes of music, arbitrarily subdivided the spectrum in these seven colors. In spoken language these are the color names most commonly used, therefore we think it's useful to identify them here by their wavelength.

Color name	Wavelength range
violet	400-420 nm
indigo	420-440 nm
blue	440-480 nm
green	480-570 nm
yellow	570-585 nm
orange	585-650 nm
red	650-700 nm

This could therefore be considered a "spoken language spectrum". In reality the spectrum is a continuum, and a range of wavelengths is associated with each of these colors, and a myriad of visibly distinct colors. Some of these, e.g. cyan (470-500 nm) do not normally appear in the "spoken language spectrum", but are very evident when one looks at the continuum spectrum dispersed by a prism.

POLARIZATION

Let us now describe one more phenomenon that can only be explained taking into account the wave nature of light. When light propagates in one direction, the electric and magnetic fields oscillate in a second and third direction, respectively, all perpendicular to each other. The plane in which the electric field oscillates, which also contains the direction of propagation, is called the *plane of polarization*. See Figures 1.7 and 1.8. The latter in particular will clarify how the direction of propagation is contained in the plane of polarization. For a simple example, imagine a rope fixed at one end and being moved at the other end. If the free end of the string is moved up and down periodically, the crests and hollows that form on the string are vertically above and below the position of the undisturbed string. In this case we can say that the waves on the string are *vertically polarized*. If the end of the string is moved horizontally, left to right, the waves are *horizontally polarized*.

To produce an electromagnetic wave, an electric charge is moved up and down on a conductor (antenna). This generates a vertically polarized electromagnetic wave, irradiating in all horizontal directions away from the antenna transmitting the signal. This is easily done by making an electric current flow back and forth, along a vertical wire.

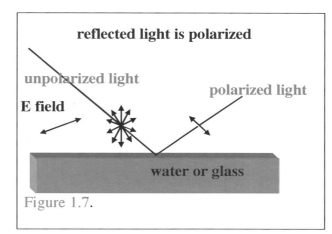

reflected light is polarized

unpolarized light

E field

polarized light

water or glass

Figure 1.7.

Light from the sun or a light bulb, however, is composed of many waves with different, random planes of polarization. This is called *unpolarized light*.

Sunlight is unpolarized: it contains electric fields oscillating in all directions. When unpolarized light is *reflected by a horizontal surface* (such as a mirror, water surface, car top, road, etc.) the *reflected light is horizontally polarized*[2], as shown in Figure 1.7.

Polaroid filters absorb light polarized in one specific direction and transmit the rest. These filters are prepared by depositing iodine needle crystals onto a polymer surface, which is then stretched so that the iodine needles orient themselves along the direction of stretching, causing them to absorb light polarized in that direction only.

Sunglasses with polarizer lenses have the iodine needles oriented horizontally and they absorb horizontally polarized light (light with the electric field E vibrating horizontally, shown in red in Figure 1.8). Vertically polarized light is still transmitted. If you combine two Polaroid filters at 90° from each other, no light is transmitted.

A *polarizer filter* in front of a camera lens darkens the clouds in the sky but darkens the sky even more. This

Sunglasses with polarizing lenses

light waves with E vibrating horizontally

light waves with E vibrating vertically

Figure 1.8

happens because light scattered by smaller particles is more polarized than light scattered by larger particles. The air molecules in the sky are smaller than the ice crystals and water droplets in the clouds. Therefore there is more polarized light from the sky than from the

---

[2] At a specific incidence and reflection angle, called the Brewster's angle, 100% of the reflected light is polarized. In other words, if light strikes the surface of a horizontal medium at a Brewster's angle, the reflected light is completely horizontally polarized. For water the Brewster's angle is 53°, for glass it is 56°, for diamond 67.5°. At incidence angles larger or smaller, part of the light is also vertically polarized, therefore the degree of horizontal polarization is less than 100%.

clouds. Correspondingly, the polarizer filter removes more light from the sky and the contrast between the clouds and the sky is enhanced, as shown in Figure 1.9.

Figure 1.9. Photographs without (left) and with (right) polarizer filter of the same clouds. Notice how the contrast cloud-sky is enhanced by the polarizer. Photographs are courtesy of Tiziana Parasassi.

Chapter 2
REFLECTION and REFRACTION

SPECULAR REFLECTION OF LIGHT

Imagine a mirror surface that is perfectly flat, polished, and reflecting. Now imagine a line perpendicular to the mirror surface, called the *normal*. The normal forms angles of 90 degrees from the mirror surface. When light illuminates any mirror like the one you imagined, it is reflected.

In Figure 2.1, the ray of light coming from the left is incident (or "incoming") on the mirror surface. The angle marked **i** is called the *angle of incidence*, and it is measured from the normal (in the particular diagram of Figure 2.1, i = 45°). The incident ray is reflected by the mirror into another ray. This reflected ray of light forms an *angle of reflection,* *marked* **r,** with respect to the normal. *The angle of incidence is equal to the angle of reflection*, or

$$i = r.$$

This is called the law of specular reflection, from the Latin word "speculum", meaning mirror. The law of specular reflection is valid for any value of the angle of incidence **i**, not only for the 45° angle reported in Figure 2.1. This law is all you really need to know about specular reflection.

Specular Reflection: $i = r$

normal

i r

i = angle of incidence
r = angle of reflection

Figure 2.1. A ray of light is reflected by a mirror, so that the angles of incidence and reflection are equal (**i = r**).

All polished and smooth surfaces reflect light as a mirror. Usually mirrors are made combining a glass pane, which can be made very flat and smooth, with a reflective metal layer coated on its back.

All materials reflect light, at least to some extent. This is true even for the most transparent ones, such as uncoated glass. You can see this in everyday life. When you are standing outside in daylight and look at a building, all the windows reflect the outside world, including the sky, other buildings and trees. Why might this happen?

This happens because the window panes are very flat and slightly reflecting. Since there is more light outside than inside the building, the windows reflect the outside world exactly as the glass reflected the candle in the previous example. The same phenomenon happens at night in the opposite direction. When there is artificial illumination inside (the lights are on) and it is dark outside, people from outside can see the inside through the window panes very well, but from the inside, people can't see outside. If you are standing inside, you can only see the reflection of yourself and the rest of the room. If you want to see outside, you must turn the inside lights off. The general rule is that you can see from darker to brighter through a window pane but not vice versa. Remember, this is because of reflection. You only see reflection if you stand on the bright side.

Another observation is that the reflectivity of polished surfaces is greater at grazing incidence, that is, when the angle of incidence (as well as the angle of reflection) is close to 90°.

Most objects around us are not perfectly polished, therefore they are not good specular reflectors. When these objects are illuminated, they do not reflect light according to the law of reflection. Instead, non-polished objects *scatter* (or *diffusely reflect*) light in all directions around them. This means that you do not have to be in a specific position or at a specific angle to see an illuminated object. You can be at any angle from it and still see it! This is quite convenient!

Every non-polished, illuminated object can be considered a source of light rays. The sun, incandescent light bulbs, neon lights, fluorescent lights, or computer monitors all emit light, but not very many other objects do. Illuminated non-polished objects, however, can be represented as sources of light, since they *scatter* light in all directions. From the geometric point of view, objects can be schematically represented as centers of diverging rays, i.e. sources of light rays. Imagine for example a red laser pointer beam, illuminating a a spot on a sheet of paper. The red dot on paper can be seen from everywhere else in the room, not only at a specific angle, as it would on a mirror. Therefore, light rays diverge from the red dot on paper as if that dot was emitting light. Clearly the paper does not emit light, it just scatters it, as most other objects surrounding us do. The dot on the paper, however, can be considered as a source of light rays, propagating in all directions.

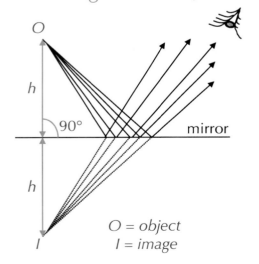

The image is **virtual**, not real

O

h

90°   mirror

h

O = object

I = image

I

Figure 2.2. Reflection from a mirror. Several observations can be made in this figure:

- Reflected rays seem to come from the point I;
- Point I is located behingd the plane of the mirror, at a distance h, equal to the distance between O and the mirror;
- Point O and point I are on a line perpendicular to the plane of the mirror;
- The mirror does not need to extend to the red line on the left: a smaller mirror is sufficient to produce an image of the object O.

Imagine now an object in front of a mirror. The diverging rays of light that originate from the object, marked O in Figure 2.2, are reflected by the mirror. The reflected rays of light *behave as if* they originated from the point I. This point is called the *virtual image* of the object O.

To find the image position of any object, draw the normal from the object to the mirror and then measure the distance between the object and the mirror. The image I is located at the same distance from the mirror as the object but it is on the opposite side. This is why it is called a *virtual image*. In other words, *there are no real rays of light that diverge from the point I. As you can see in Figure 2.2, only the artificial extensions of the reflected rays (marked as dashed lines) originate from I.*

Any ray from O that strikes the mirror is reflected and the artificial extensions of these reflected rays always seem to come from point I. If you were to remove the mirror, and place the point object O at the point I, it will look identical to the reflected point O. The eye cannot tell the difference between rays coming from a virtual image I and a real object at the same position.

When you look at yourself in the mirror, what you see is a virtual image of yourself. This virtual image is behind the mirror. If a friend of yours were to look behind the mirror, however, they would not find an image of you. You see: the image is virtual!

You can only see a virtual image if you are on the same side of the mirror as the object.

Another important observation is that vertically mounted mirrors appear to invert left and right, but not top and bottom. Both these observations are actually incorrect. The vertical mirror does not invert left-to-right, it only inverts front and back. In the mirror on the wall you see the top of your face on the topand your right ear on the right. Neither up-down nor left-right are reversed. The only inverted direction is front and back. Other people facing you, however, always see your right ear on their left, thus they see you differently from the way you see yourself in the mirror. This is why everybody thinks that left and right are inverted in a vertical mirror. It's just the habit we all have of seeing each other frontally.

Why is gold yellow? Why are silver and aluminum gray? Why is copper red?

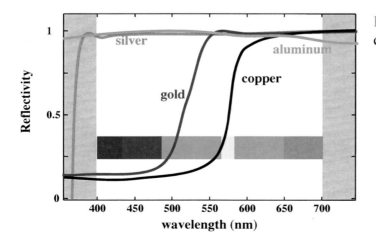

Figure 2.3. Reflectivity curves of several metals.

In Figure 2.3 we report the *reflectivity curves* for these metals. As mentioned in Chapter 1, the wavelengths between 400 nm and 700 nm are the *visible range* of light. Below 400 nm is ultraviolet light and above 700 nm is infrared light. We do not see these types of light. From the reflectivity curves, you can see that gold does not reflect blue. Rather, it absorbs blue and this causes it to appear yellow. Copper absorbs both blue and green and therefore appears orange-red. Aluminum and silver have fairly flat reflectivity curves along the entire visible range so they appear gray. In other words, aluminum and silver reflect all the colors of light illuminating them and absorb none. This is why silver and aluminum are the best materials for mirrors. In the old days, silver was the only material used to make mirrors, and that's why mirrors were so few and expensive. Today, technology is available that allows us to deposit thin smooth layers of aluminum on glass. Mirrors made through this process are much more affordable, so now mirrors are much more widely available. Can you imagine life without a mirror? If not, you may be thankful to physics!

REFRACTION OF LIGHT

Light travels along straight lines in a uniform medium. If a ray of light passes from one medium to another (from air into water or glass, for instance), in a direction that is normal, or perpendicular, to the interface, the ray will travel straight. However, if a ray passes from one medium to another *in a direction different from the normal*, the direction of the light ray changes in the second medium—*the ray is kinked*. This abrupt deflection, or kinking, of light rays at the interface between two media is called *refraction*.

The behavior of light in a medium depends on the *index of refraction* of the medium, n. The index of refraction of a medium determines the speed of light in that medium and the angle or refraction. The index of refraction n is an intrinsic property of each material, and depends on its density. In general, the denser the medium the higher the index of refraction n. Vacuum, and, to a good approximation, air have an index of refraction n = 1. For water, n = 1.33, and for glass, n = 1.52. The index of refraction n is a dimensionless number, in other words it has no units. In most media, n is a number between 1.0 and 2.5. A table of the indices of refraction of common materials is shown below.

material	index of refraction n
vacuum	1.00000
air at 50°C	1.00024
air at 0°C	1.00029
liquid water (20°C)	1.33
ice	1.31
ethanol	1.36
plexiglass	1.50
glass (crown)	1.52
glass (flint)	1.89
zircon	1.97
cubic zirconia	2.16
diamond	2.42
gallium phosphide	3.50
gallium arsenide	3.93
silicon	4.01

In reality, things do not work out so simply. There is an additional complication: the index of refraction is not constant for each material but actually depends slightly on the wavelength of light. This phenomenon is called *dispersion*. For example, the index of refraction of borosilicate crown glass (known also as BK7) for short 400-nm wavelength (violet) is $n_{400}=1.53$ and for long 700-

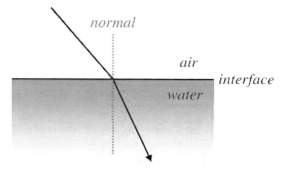

Figure 2.4. Light traveling from air to water: when it traverses the water surface, the light ray is kinked towards the normal.

nm wavelength (red) is $n_{700}=1.51$. This is may seem a small difference, but, as we will describe later on, it makes it possible to separate the spectral colors using a crown glass prism.

Let us consider a light ray crossing the interface between a medium with an index of refraction n = 1.0 and a second medium with an index of refraction n > 1. This is depicted in the diagram below. Notice in the diagram that a ray of light is traveling in air and then reaches the interface between air and water. As the ray crosses the interface and enters the water, it is kinked towards the normal. Whenever light travels from a less dense to a denser medium, it kinks *towards the normal*, as shown in Figures 2.4 and 2.5. When light travels in the reverse direction from a denser to a less dense medium it kinks *away from the normal*.

Figure 2.5. Definitions of the angles of incidence and refraction.

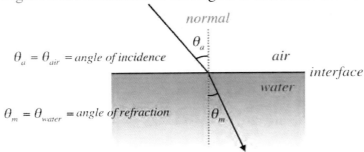

Let us compare the angle that a ray forms with the normal at the point of incidence (θ_a, the angle in air) and the corresponding angle in water (θ_m), shown in Figure 2.5. The behavior of the ray as it is refracted at the interface can be described as follows:

a. The angle in air (θ_a) and the angle in the medium are on opposite sides of the normal (Figure 2.5).

b. The angle in air is always larger than the angle in the other medium because air is less dense than any other media (e.g. glass, water, plexiglass, acrylic, etc..)

c. The angles θ_a and θ_m do not change if the light ray goes from air to the medium or from the medium to air. The ray of light is reversible. This is always true. *Light rays are always reversible*. We will make great use of this characteristic of all light rays, here and later.

d. The angles θ_a and θ_m are related to each other in a very accurate and quantitative law, called Snell's law, shown below. Figure 2.6 shows a visual representation of this law.

SNELL'S LAW
$$\frac{\sin\theta_a}{\sin\theta_m} = \frac{n_m}{n_a}$$

Refraction: Simplified Snell's Law

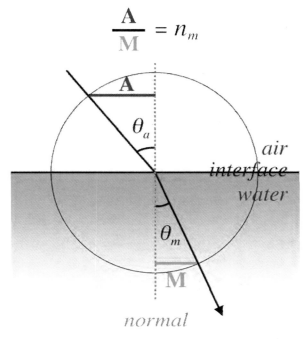

$$\frac{A}{M} = n_m$$

Figure 2.6. A visual representation of Snell's law, in which A is the sine of θ_a and M is the sine of θ_m. These correspond to the blue and green horizontal lines, respectively.

TOTAL INTERNAL REFLECTION

Any angle in air θ_a corresponds to a particular angle θ_m in the medium. For example, for rays traveling from air into water (with an index of refraction n = 1.33), Snell's law gives the following results:

angle in air θ_a	angle in water $\theta_m = \sin^{-1}(\theta_a/n)$	
0°	0°	
10°	7.5°	
20°	14.9°	
30°	22.1°	
40°	28.9°	
50°	35.2°	50°
60°	40.6°	
70°	44.9°	35.2°
80°	47.8°	
90°	48.7°	

The table goes up to 90° only. This is because the angle in air, θ_a, can never be greater than 90°. If it were greater, the ray would no longer be coming from air. It would be coming from under the surface of water (below the interface).

The table can be also used for a ray of light traveling in the opposite direction, from water into air. A diver with a waterproof flashlight could experiment with rays of light and verify for us that the angles reported in the table are still accurate when going from water to air.

Remember, *rays of light can always be inverted*. All optics formulas and all drawings, including Snell's law and the table of values for θ_a and θ_m, do apply to inverted rays. When using inverted rays, the only difference for Snell's law is that you start from the angle in water and calculate the angle in air. Looking at the table on the previous page, you would start from the right column and read the corresponding angles on the left. There is one thing you must watch out for when using inverted rays. Since $n_{water} > n_{air}$, the angles in water θ_m do not go all the way to 90°. Notice: the table does not tell us what happens with a ray with $\theta_m = 60°$. What happens is simply the following:

$\theta_m = 49°$ corresponds to $\theta_a = 90°$

$\theta_m = 60°$ corresponds to $\theta_a > 90°$. This is *impossible* because this ray would not really be in air. The ray is not refracted into air but *reflected* at the interface of water-air back into water!

Above a certain angle, called the *critical angle* θ_c or, more precisely, the critical *angle of total internal reflection*, all light is reflected from the interface. No light is refracted. This occurs only when the ray of light passes from a denser to a less dense medium, that is, from higher to lower index of refraction. For example, this can happen when a ray is passing from water to air, as described above, but not from air to water.

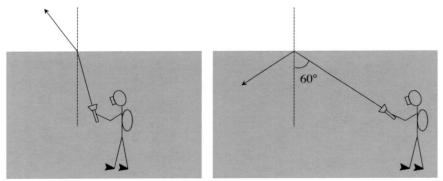

Figure 2.7. A man under water can project light into air from his flashlight, only as long as the incidence angle is less than 49°, the critical angle of total internal reflection for the water/air interface.

The critical angle for the water/air interface is 49°. For glass/air it is 41° and for diamond it is 24.4°.

This means that if the angle of incidence is greater than the critical angle, $\theta_m > \theta_c$, *no light is refracted*. In Figure 2.7, since the angle of incidence is 60°, and it is greater than the critical angle $\theta_c=49°$, the light is totally reflected.

It is easy to calculate the critical angle. At the critical angle, the angle in air $\theta_a = 90°$. Since we know that the index of refraction n for air $=1$ and that $\sin 90° = 1$, Snell's Law appears as: $n_{water} = \dfrac{1}{\sin \theta_{c(water)}}$

Hence, $\sin \theta_{c(water)} = \dfrac{1}{n_{water}}$

from which we can calculate $\theta_{c(water)} = \sin^{-1}(\dfrac{1}{n_{water}}) = \sin^{-1}(0.75) = 49°$.

With the aid of the above information, we can anticipate the behavior of a light ray as it crosses a transparent medium. This allows us to make drawings of what will happen without actually doing experiments! In Figures 2.8, 2.9, 2.10 and 2.11 we describe a few examples of refraction in some common situations. In all of these examples, for simplicity's sake, we only show the rays coming from one point of the object. Be aware that there are similar rays coming from each point on the surface of the object. You can imagine what these rays would look like.

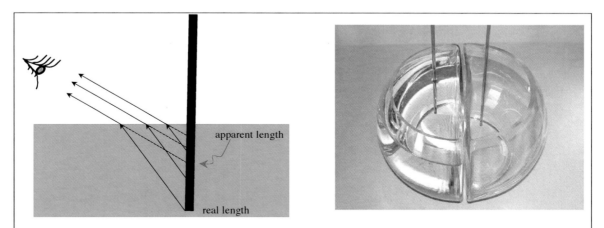

Figure 2.8. Because of refraction, a vertical stick under water appears to be shorter. If the stick were not vertical, it would also appear to be bent at the air-water interface. The bottom of the stick always appears closer to the surface than it really is. The photograph shows two identical bowls (approximately semispherical) and two turquoise glass rods, which are vertical and identically positioned in the two bowls. The left bowl if filled with water, while the right bowl is not. Notice the striking difference in apparent length: the under water rod appears shortened by refraction.

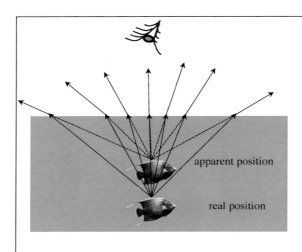

Figure 2.9. A fish underwater, seen from the top, appears to be closer to the surface than it is in reality. The apparent size of the fish, however, does not change.

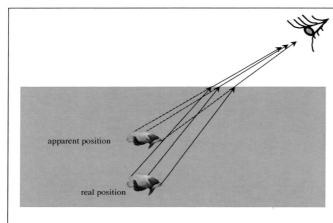

Figure 2.10. A fish seen at an angle (not from the top) appears to be closer to the surface and more vertically compressed (narrower) than it really is. Experienced spear-fishermen keep refraction into account when they aim and launch their spear.

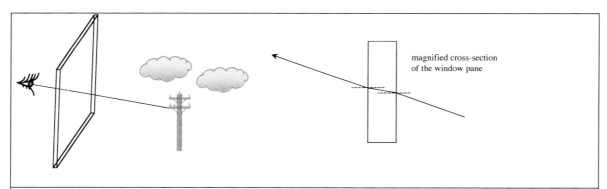

Figure 2.11. Here a ray crosses a glass pane with parallel sides. The ray coming from the right is kinked towards the normal (dashed line) when entering the glass, and away from the normal when exiting on the left. The incoming and outgoing rays are parallel to each other. Since the window pane is thin, there is almost no shift in the image position. A thicker block of glass or plexiglass shifts light rays much more than a window pane.

REFLECTION AND REFRACTION IN DIAMONDS

A brilliant cut diamond sparkles brilliantly because of skillful use of refraction and total internal reflection. Let us look at the detailed path of some rays of light. Consider two rays of light (1 and 2 in Figure 2.12) passing the air-diamond interface. Ray 1 is kinked towards the normal (red), because a diamond is denser than air. Ray 2 enters perpendicular to the surface of the diamond (along the normal), and therefore does not kink. Instead, it goes straight through the surface. Both rays reach the inner surface at the arrow heads, and from there, undergo total internal reflection. The angles that fall within the critical angle are shown by the blue cones in Figure 2.12. Both rays 1 and 2 are *outside* these blue cones and are thus both larger than the critical angle. Both rays are, therefore, totally internally reflected. The critical angle is calculated, as before, using Snell's Law. The index of refraction for diamond is $n_{\text{diamond}} = 2.42$.

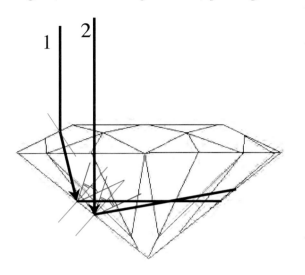

Figure 2.12. Rays 1 and 2 are totally internally reflected, because they reach the inner diamond surface at angles greater than the critical angle, which for diamond is very small: only 24.4°.

Snell's law for diamond is:

$$\frac{\sin \theta_{air}}{\sin \theta_{diamond}} = \frac{n_{diamond}}{n_{air}}$$

Total internal reflection in diamond takes place when $\theta_{air} = 90°$, which corresponds to

$$\sin \theta_{diamond} = \frac{n_{air}}{n_{diamond}} = \frac{1}{2.42} = 0.41 \text{ and } \sin^{-1} 0.41 = 24.4°$$

Whenever $\theta_{diamond} - 24.4°$, light is completely reflected within the diamond and no light is refracted outside.

After rays 1 and 2 are totally internally reflected, the two rays reach the other side of the diamond bottom surface. Here, they are again both traveling at angles greater than the

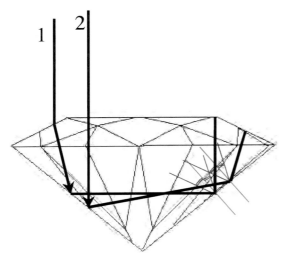

Figure 2.13. Rays 1 and 2 are again totally internally reflected towards the top of the diamond.

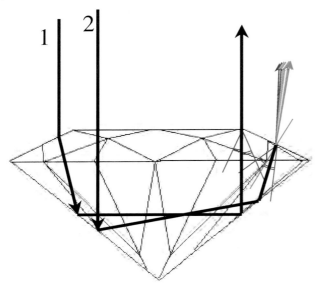

Figure 2.14. Light of different colors is dispersed at different angles, generating the "fire" in a diamond.

critical angle (outside the blue cones), as shown in Figure 2.13. Therefore, they are completely reflected towards the top of the diamond.

Ray 1 is reflected in a vertical direction and reaches the top surface perpendicularly. Therefore, it goes straight through the surface. Ray 2 reaches the surface at an angle. This time, the incidence angle is *within* the blue cone, so the ray is refracted outside the diamond. Since the index of refraction n is slightly different for each light wavelength, each color of the spectrum is refracted at a slightly different angle as shown in Figure 2.14. Violet rays are kinked more than red rays. In physics, this phenomenon is called *dispersion of light*. In the jewel industry, it is called the *fire* of a gemstone. Light dispersion will be discussed in more detail in Chapter 6.

If the diamond is cut too shallow or too deep (as shown in Figure 2.15), light leaks out through the bottom facets and the diamond appears dark and dull. The cones defining the critical angle are again shown in blue. Notice where the light leaks out, the rays are traveling at an angle that falls within the blue cones. If light is able to "escape" in this way, a diamond does not appear as sparkly.

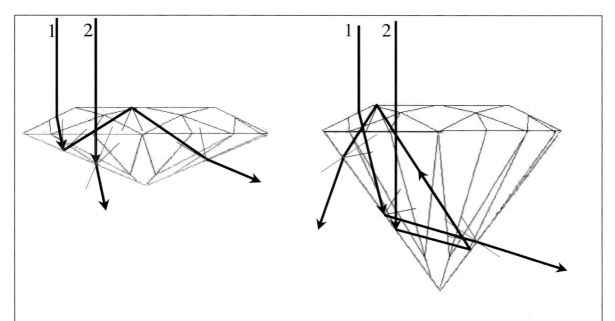

Figure 2.15. Two diamonds, one cut too shallow and one too deep, are shown here. The cones defining the critical angles are again shown in blue. Notice where the light leaks out, the rays are traveling at an angle that falls within the blue cones.

For the diamond to appear sparkly, it must be cut at very accurate proportions, for instance the "brilliant cut" and its proportions, optimized to give the greatest sparkle to a diamond, are shown in Figures 2.16 and 2.17.

proportions of a brilliant cut diamond

Figure 2.16. The precise proportions of the "brilliant cut" for diamonds.

Figure 2.17. Top view of the circular crown (top) and side view of a brilliant cut diamond (bottom).

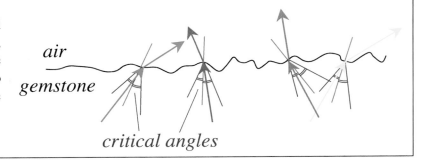

Figure 2.18. Diamond facets must be polished, otherwise most of the light is lost, due to refraction within the critical angle cones.

After cutting the gem, the surface of the diamond facets must also be polished. If they are not polished, the angles of total internal reflection are not all oriented equally with respect to the surface, as shown in Figure 2.18. This causes most of the light to be lost to refraction.

THE RAINBOW

Now that we have familiarized ourselves with reflection, refraction and dispersion, we can understand how rainbows form. In general, to be able to see a rainbow in front of you, the sun must be behind you. A myriad of water droplets, such as rain or moisture, must be present in the air for the rainbow to appear. This is shown in Figure 2.19. Three of the droplets are magnified there. The black lines represent light rays coming from the sun. Droplet 1 shows how the white light from the sun is *refracted* inside the droplet (toward the normal) and then *dispersed* so that violet rays are deflected more than red rays. All of the light rays are then partly *reflected* on the inner surface of the droplet, following the law of reflection. Notice that this is *not* total internal reflection! In fact, light rays are partly reflected and partly refracted outside the droplet (away from the normal). The reflected rays then reach the droplet surface again and, again, will be partly reflected (see droplet 1 in Figure 2.19) and partly refracted outside the droplet.

Droplets 2 and 3 show the red and the violet light rays only. Hopefully this will clarify that there are different angles and directions of refraction and internal reflection for each color wavelength. In reality, all colors come out of all droplets at all times, so it is incorrect to assume that 2 is a "red droplet" and that 3 is a "violet droplet". They only appear to have those colors to the little man with the umbrella because these droplets are at a well-defined position in the sky. This

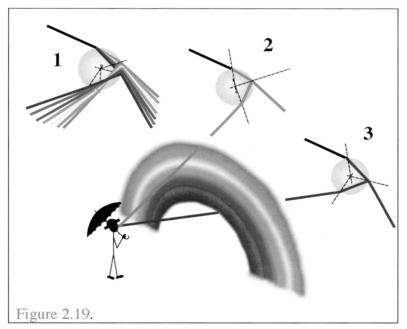

Figure 2.19.

allows the refracted-reflected-refracted rays reach his eyes at the correct angle to appear red and violet, respectively. Notice how the violet rays come out of the droplets at a shallower angle while the red rays come out at a steeper angle. Likewise, red always lies along the outside arches of the rainbow while violet is always along the inside.

A straight line from the sun through the eye of the man with the umbrella and continuing on marks the center of the rainbow. In this sense the man with the umbrella carries the rainbow around with him. Sometimes on top of a mountain or from an airplane one can see almost an entire rainbow circle.

On rare occasions *double rainbows* can be observed in nature. The second rainbow always lies outside the first one and always has an inverted color order (violet outside, red inside). The second rainbow is formed by the second internal reflection and the fourth refraction outside of the droplet that follows. The droplet in Figure 2.20 shows a schematic representation of all reflections and refractions.

Figure 2.20.

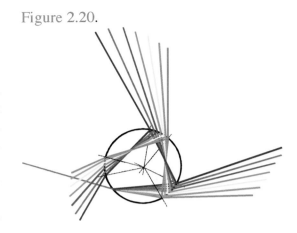

Notice that, at each refraction, some of the light is lost in a direction that cannot be observed from ground. Therefore, the second rainbow is always dimmer than the first. At this point it may be a bit difficult to imagine how these two rainbows can form at once. Figure 2.21 shows a droplet that could "potentially" generate both rainbows. In reality, no droplet alone can generate either a single or a double rainbow. You have to imagine a myriad of droplets.

Figure 2.21.

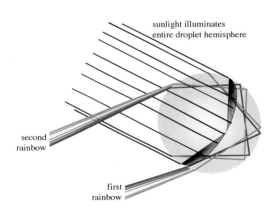

sunlight illuminates
entire droplet hemisphere

second
rainbow

first
rainbow

Those droplets that are at the correct position for an observer to see them as red will appear red. The next ones will appear orange, the next yellow, the next green, and so forth. Observe how the doubly-reflected rays that form the second rainbow have an inverted color order with respect to the first rainbow. Inside the droplet of Figure 2.21, again for simplicity's sake, only violet and red rays are shown. The final refracted rays (the rainbows), however, are depicted as the seven commonly mentioned colors of the spectrum. Be aware that this is only a simplification. There are not really only seven colors in a rainbow but an infinite number of colors all associated with the infinity of continuously increasing wavelengths along the entire visible light spectrum. Of this infinity, the human eye can distinguish approximately 10 million colors. Water droplets disperse all different wavelengths with continuity, along with the seven colors that are commonly associated with a rainbow. No abrupt separations between colors can actually be observed in the rainbow or any other dispersed spectrum.

Chapter 3
LENSES

THE PRISM

When the surfaces of a block of glass or plexiglass are not parallel, as in a prism, the incoming and outgoing rays of light are not parallel. In Figure 3.1, the incoming ray is first refracted and kinked towards the normal 1. Then it is transmitted through the prism and when it reaches the other surface, it is refracted again, but this time away from the normal 2. When light travels through a prism, the end result is *always* that the ray is deflected *away from the apex* of the prism.

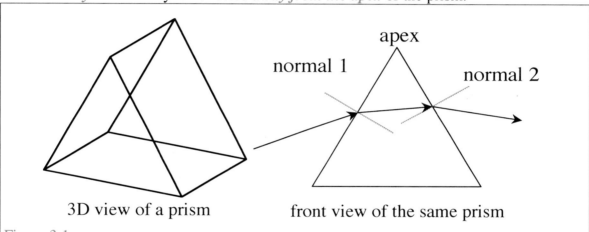

apex

normal 1

normal 2

3D view of a prism

front view of the same prism

Figure 3.1.

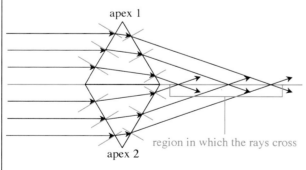

apex 1

region in which the rays cross

apex 2

Figure 3.2. Combination of two prisms. The top prism deflects the rays down, while the bottom prism deflects them up. The rays that travel through equivalent sections of the two prisms cross each other at the center, on the pink axis. But is there a transparent object that can make *all* parallel rays that travel through it cross at the exact same point, no matter where they traverse the object?

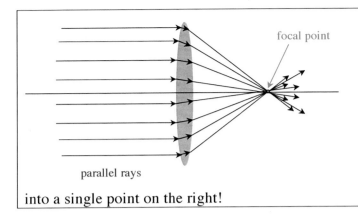

Figure 3.3. The answer to the question in Figure 3.2 is yes! If the two surfaces of the transparent prisms are not flat but curved along a spherical surface, the result is a lens. In the diagram here, the lens is convex on both sides, and therefore it is a converging lens. All parallel rays coming from the left *converge* into a single point on the right!

CONVERGING AND DIVERGING LENSES

Converging lenses are *thicker* in the middle than near the edges and so they *deflect light rays towards an axis*. You can see this if you apply the law of refraction to the surface where the light beam enters the glass and to the surface where the light beam comes out of the glass to find the direction in which light will kink with respect to the normal). You may think of a little piece of the lens as similar to a prism. Figure 3.4 shows a few possible light rays.

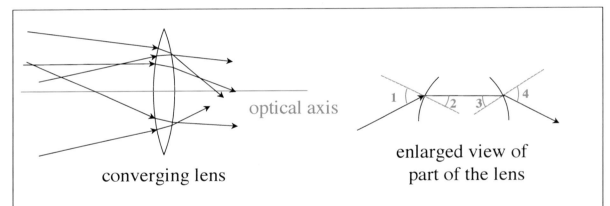

Figure 3.4. Notice in the enlargement that, in accordance with Snell's Law, angle 2 is smaller than angle 1, and angle 4 is larger than angle 3.

Figure 3.5. *Diverging lenses* are *thinner in the middle* and so they deflect light rays *away* from the axis.

optical axis

diverging lens

Some other lens shapes:

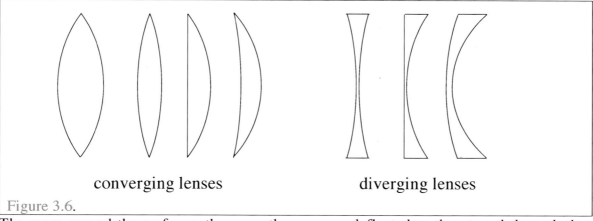

converging lenses   diverging lenses

Figure 3.6.

The more curved the surfaces, the more the rays are deflected as they travel through the lens. The higher the refractive index of the lens material, the more the rays are deflected.

Eye glasses are diverging or converging depending on whether the person is nearsighted or farsighted. If you wear glasses, check to see if they are thicker or thinner at the outer edges.

FOCAL LENGTH
Consider parallel light rays (for instance light rays from a light source very far away, or at infinity) falling upon a converging lens.

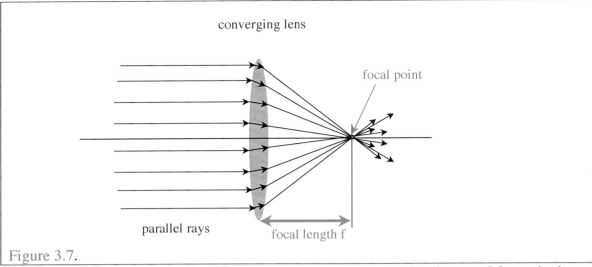

converging lens

focal point

parallel rays

focal length f

Figure 3.7.

As shown in Figure 3.7, the rays converge at a point at a certain distance **f** from the lens. This is called *the focal length* of the lens. Conversely, if you place an object (a source of light) at distance f from the lens, the light rays will pass through the lens and come out parallel to each other (light rays can always be inverted). A *strong lens* is one that deflects light a lot, meaning it has a *short focal length*. The focal length of camera lenses is always given in millimeters. The most common and inexpensive 35 mm cameras (indicating 35 mm film width, *not* 35 mm focal length!) have lenses with a focal length of 50 mm (f = 50 mm). For eyeglasses, the *dioptric power* is usually used instead of the focal length to describe the lens. The relationship between dioptric power (measured in diopters, D) and f is simple:

$$\text{Dioptric Power (D)} = \frac{1}{f\,(m)}$$

The focal length *f* in this formula *must* be expressed in *meters*! If you forget, it is like mislabeling a piece of mail. If you put the wrong house address, the letter may get to the right street— but a few blocks away. Similarly, the diopters will be off by several decimal places. In physics, that's a big difference!

According to the formula, a camera lens with a focal length f = 50 mm = 0.05 m has $\frac{1}{0.05}$ = 20 D. The advantage of diopters is that if you combine two lenses, one right next to the other, you can simply add the diopters. This implies that for two lenses, the combined focal length is:

$$\frac{1}{f} = \frac{1}{f_1} + \frac{1}{f_2}$$

where f_1 and f_2 are the focal lengths of the two lenses.

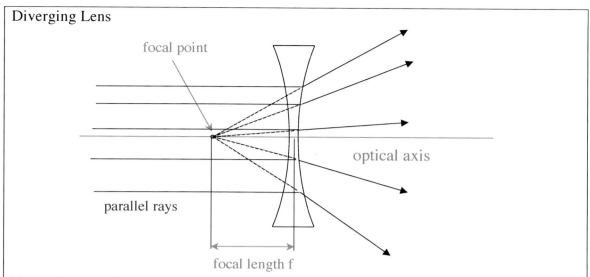

Diverging Lens

focal point

optical axis

parallel rays

focal length f

Figure 3.8. A diverging lens with parallel rays incident upon it from a far away object. The diverging lens makes the rays diverge, as if they were originating from the focal point. The image of the far away object (at infinity) on the left is formed in the focal point on the left—the same side of the lens as the object. *The image is virtual,* as there are no actual rays converging at the image point. Only the artificial extension of diverging rays (shown as dashed lines in this diagram) converge in the focal point, or, more accurately, *diverge from the focal point.*

After the parallel rays go through the diverging lens, they diverge *as if* they came from the virtual image at the focal point. Just like with converging lenses, the focal point is found at a distance f, or the focal length, away from the lens. To indicate that the (apparent) crossing point of the rays is on the same side of the lens as the source of light, *the focal length of a diverging lens is stated as a negative number.* An important advantage of giving the focal length of a diverging lens as a negative number is that it makes the formulas work out automatically. For instance, if you combine a converging lens with 10 cm focal length with a diverging lens of -10 cm focal length the resulting focal length f is calculated from the following equation:

$$\frac{1}{f} = \frac{1}{10} + \left(\frac{1}{-10}\right) = \frac{1}{10} - \frac{1}{10} = 0$$

$f = \infty$

Figure 3.9.

In the system of two lenses shown in Figure 3.9, the resulting focal lens is infinite—the incoming rays are parallel, the outgoing rays are also parallel. The lens combination acts like a flat window pane! But wait: does this really work? Yes, because when a converging lens and a diverging lens are combined so that the convex and concave sides are glued together, the resulting object is indeed a parallel slab of glass, just like a window pane!

The focal length of a lens depends on the index of refraction of the glass and on the radius of curvature of the two faces of the lens, r_1 and r_2:

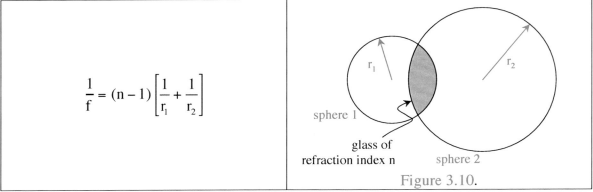

$$\frac{1}{f} = (n-1)\left[\frac{1}{r_1} + \frac{1}{r_2}\right]$$

sphere 1

glass of refraction index n   sphere 2

Figure 3.10.

This is called the *lens-maker's formula*. This formula has important consequences that are worth noticing:

There are always two focal points for each lens, and they lie at a distance f on either side of the lens.

The lens may have two different radii of curvature, but the focal points are always equidistant from the center point of the lens[1]. This implies that a lens can be used from one side or flipped and used from the other side, and either way the image will be identical. If you wear glasses, you can try this right now! Take your glasses off, rotate them by 180° so that you can look through the lenses and the arms extend away from your face. Now look through the lenses. If you rotated the glasses correctly, you should be looking with your right eye through the right lens and with the left eye through the left lens, but inside-out. You should be able to see just as well with the lenses inverted. Taking into account how the lenses are curved, this is really counterintuitive! It is one of many surprising phenomena in physics.

The formula makes sense even if n = 1 (a lens made of air!), $\frac{1}{f}$ is always zero or f→∞ and the rays are not deflected.

If the face of the lens is *concave*, r is counted *negative*.

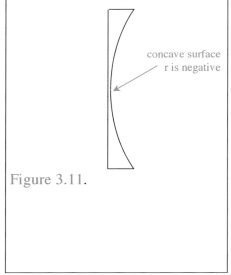

concave surface
r is negative

Figure 3.11.

The smaller r_1 and r_2 get, the smaller the focal length. What does the formula tell you if you take a thin sheet of plastic and curve it? Does it still make sense?

Note that, the lens-maker's formula *works only for thin lenses*—lenses for which the thickness is much less than the focal length.

---

[1] This is always true only for thin lenses.

IMAGES - REAL AND VIRTUAL

Imagine sunlight shining on the face of a lion. Think of one tiny point on the lion's cheek. Part of the sunlight hitting that point is *scattered in all directions -* that's why we see the lion with our eyes no matter from what direction we look at it! This is also called diffuse reflection of light, and is different from specular reflection from a mirror, which goes in one particular direction only. If a converging lens is placed in such a

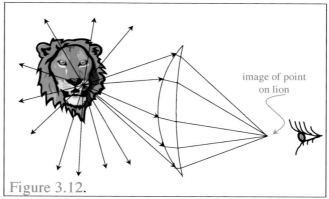

image of point on lion

Figure 3.12.

way that the rays coming from the point of the lion converge to a point on the other side of the lens, an image of the object point on the lion is formed. Figure 3.12 shows this case. The lion is the object, and for each point on the lion an image point is formed on the other side of the lens. If you put a screen at the image point, you see an *image* of the point on the lion. This is a real image of the object: there are real light rays converging at the image point. Therefore, if you put a piece of paper at the image point, the light rays will form a focused spot on the paper. In fact, you will see an image of the whole lion because, point by point, the rays traveling from the lion converge again on the other side of the lens to form a full image of the lion, as shown in Figure 3.13.

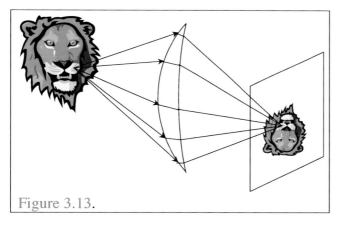

Figure 3.13.

If there is no screen, you will see the image only if you stand on the back of the lens. From the side you will not see the image of the lion. The best way to see the image is to place a screen where the image is formed, as shown in Figure 3.13, because the screen will direct some of the light into your eyes by, again, the process of scattering. The image is called *real* if (after passing through the lens) *the rays* from a given point of the object *cross each other, or converge, into a point.* A real image can be seen on a screen or on a photographic film. To form a real image we need a converging lens, but *converging lenses do not necessarily produce a real image,* as we shall see later.

The distance from the object to the lens is called the object distance, represented as o. The distance from the lens to the image is called the image distance, or i. It is important to always remember that the distance i may be where the rays from the object are focused but this is *not* the focal length!!! Focal length refers to the place where **parallel incident** rays converge—this *requires* that the object be very far away!! The image distance i is equal to the focal length f if and *only if* the object distance o is very large or at infinity (written as the symbol ∞).

If the object is very near the converging lens, the lens may not be strong enough to deflect the light rays coming from the object and make them converge into a point. Worse yet, we may have a diverging lens that spreads the rays even more than they were spreading before they got to the lens. In either case, after crossing the lens, the rays *seem* to come from a point at distance i behind the lens (with respect to your eye, see Figures 3.14 and 3.15). You can see the image with your eye if you look through the lens, but you cannot place a screen or a film where the image is, and expect to see an image—no real rays cross in that location, or anywhere else for that matter!

This is a *virtual image*. We have already encountered virtual images: in a mirror, the image produced by reflection is also virtual. If you place a light bulb near a mirror, there is no way to catch the image on a piece of white paper, no matter where you place it. To do this, you would need an additional lens (such as your eye or a camera taking a picture of the rays reflected by the mirror) to make the rays converge and form a real image. A *converging lens* forms a virtual image only when the object is closer to the lens than the focal length (Figure 3.14). A *diverging lens* always produces virtual images (Figure 3.15).

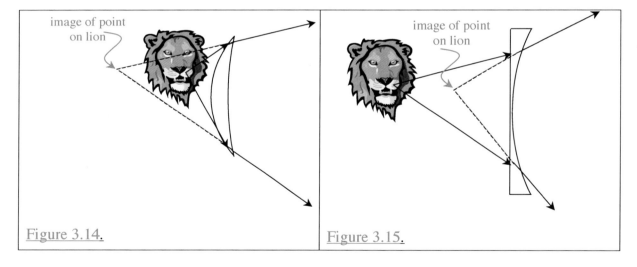

Figure 3.14.　　　　　　　　　　　　Figure 3.15.

THREE EASY RAYS

Given a thin lens and an object, one can graphically draw rays to find the image size and position. With a diagram, you can see how the image is formed, and where it is. This procedure is called *ray tracing,* and is performed by very fast computers that trace many rays of light simultaneously, to simulate what happens in nature. Ray tracing is used in scientific applications, to produce sophisticated camera lenses, in animation movies, in computer graphics, and will be used by you in this course! It may seem complicated but it is very simple to learn how to master ray tracing for a singe thin lens. The trick is to focus on three particularly easy rays.

If we select a point on the object, for instance the top of the light bulb in Figure 3.16, we do not need to repeat the operation for each and every point of that same light bulb. We only need to use three easy rays!

The *three easy rays* to be traced *for a converging lens* are shown in Figure 3.16, and described verbally hereafter. They are:

• A ray from the object *through the center* of the thin lens. This ray goes *straight through* because at the center of the lens the two lens faces are parallel to each other. There is actually a very slight shift for thicker lenses (see Figure 2.11), but it can be neglected for thin lenses like the ones we discuss here.

• A ray from the object to the lens, *parallel to the optical axis.* This ray goes *through the focal point on the other side of the lens.* The image point is where these two rays cross, so now we can draw the image as shown in the graph. It is an upside down, or inverted, image.

• A ray from the object *through the focal point* on the left[2] of the lens. This ray will leave the lens traveling *parallel to the axis* and, of course, should go through the same image point already found. This is a good way to check the accuracy of the first two rays!

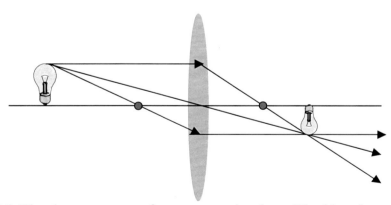

Figure 3.16. The three easy rays for a converging lens. The blue dots mark the distance f (focal length) from the lens. Remember, the distance is the same on either side of the lens.

*The three easy rays **for a diverging lens**,* shown in Figure 3.17, are:

_____

[2] The third ray does not have to go through the focal point of the left. It has to go through the focal point on *the same side of the lens as the object.* Therefore, if the object is on the right it goes through the right focal point.

- As for the converging lens, the first easy ray goes straight *through the center of the lens*.

- The second ray is *parallel to the optical axis*. This ray is deflected out of the lens and appears *as if it came from the focal point on the left*[3] side of lens. The image point is where the first ray and the *extension* (dashed line) of the second ray cross.

- The third ray is the ray from the top of the object *heading in the direction of the focal point on the right of the lens*[4]. This ray will come out *parallel to the axis*, seeming to come from the point already identified by the first two rays. This ray may seem more complicated to draw. Just remember that light rays can always be inverted, and therefore, if this third ray were to come from the right traveling parallel to the optical axis, it will diverge on the left and seem as if it came from the focal point on the right.

We see, from these three easy rays that the image is virtual, upright, and smaller than the object. This is always the case for diverging lenses.

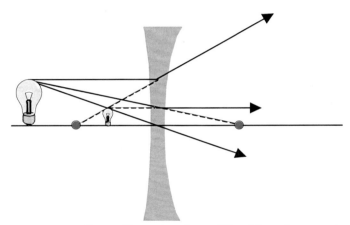

Figure 3.17. The three easy rays for a diverging lens. The blue dots again mark a distance f from the lens.

THE LENS FORMULA

The lens formula can be used to quantitatively answer a variety of questions: Where is the image? Is the image real or virtual? How large is the image? The lens formula tells you how *object distance o* and *image distance i* are related to the focal *length f* of a lens:

$$\frac{1}{o} + \frac{1}{i} = \frac{1}{f}$$

---

[3] Again, this does not have to be the left side, just the same side as the object.

[4] As for the previous notes: this is not necessarily on the right, but has to be on the opposite side of the lens with respect to the object.

Figure 3.18 explains how to measure the magnification. Given a certain object distance o and a given focal length f, one can calculate the image distance i using the lens formula. The magnification M, then is simply the ratio of i over o.

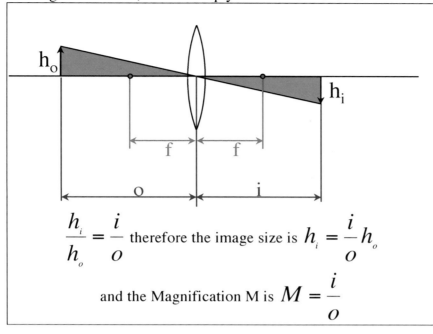

$$\frac{h_i}{h_o} = \frac{i}{o}$$ therefore the image size is $$h_i = \frac{i}{o} h_o$$

and the Magnification M is $$M = \frac{i}{o}$$

Figure 3.18. The magnification M can be easily calculated. The method is shown in this diagram. In the formulas, h_o is the height of the object and h_i is the height of the image. The formula is directly derived from the two similar triangles filled with green.

Be careful with the term magnification. Keep in mind that the magnification can be smaller than 1, and when this is the case, the lens really is *demagnifying* and the image produced is smaller than the object. The range of possible magnifications for a given converging lens varies dramatically, as the lens can be either magnifying or demagnifying depending on how far the object is positioned from the lens.

The magnification is always a positive number: if the image distance i is negative (object and image are on the same side of the lens), take the modulus of i, or the absolute value — *ignore the minus sign.*

Let us now see some examples of how the lens formula and the magnification formula are used. For the next three examples, take a converging lens with a focal length of 10 cm. Notice that the focal length is always given by the manufacturer of the lens, and cannot be changed (this is not true for zooming lenses, in which the focal length can be changed, but we are now talking about simple, single lenses).

1. Object Distance Greater Than Focal Length:
First, place the object 20 cm from the lens. Where is the image?

$$\frac{1}{i} = \frac{1}{f} - \frac{1}{o} = \frac{1}{10} - \frac{1}{20} = \frac{2}{20} - \frac{1}{20} = \frac{1}{20}$$
i = 20 cm

so, the image is 20 cm from the lens.

How big is the image?

The height of the object is h_o, and the height of the image is h_i.

$$Magnification = \frac{h_i}{h_o} = \frac{i}{o}$$

In other words, the image size is to the object size as the image distance is to the object distance—the two lengths are proportional!

In this example $\frac{h_i}{h_o} = \frac{i}{o} = \frac{20}{20} = 1$ or $h_i = h_o$.

Therefore, for this example the image is the same size as the object.

2. Using the same lens, move the object to a distance of 50 cm away:

o = 50 cm ; f = 10 cm

$$\frac{1}{i} = \frac{1}{10} - \frac{1}{50} = \frac{5}{50} - \frac{1}{50} = \frac{4}{50}$$

$$i = \frac{50}{4} cm = 12.5 cm$$

Image size is now 1/4 [=12.5/50] of object size. Note that as we moved the object from 20 to 50 cm (further away from lens), the image got closer to the lens (from 20 cm to 12.5 cm).

There is a neat and easy way to understand (and more importantly to remember) this. Look at Figure 3.19: light hitting point A of the lens is deflected as indicated by the angle labeled "deflection".

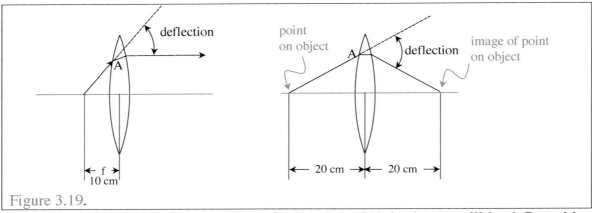

Figure 3.19.

If the object is in the focal point, (10 cm in Figure 3.19, left) the ray will be deflected by the lens in point A so that it comes out of the lens parallel to the optical axis. This deflection angle stays the same, no matter where the object is placed. So, if for example we put the object at twice the focal length (20 cm, Figure 3.19, right) its image will be formed at a distance that keeps the deflection angle the same, i.e. at exactly 20 cm. If we now move the object further away (as in Figure 3.20, left), the ray coming from the same point on the object and hitting the same point A will still be deflected by the same angle as before! The only difference is that the incoming ray hitting point A comes in at a lower angle (less steep) when the object is further away, causing the ray leaving the lens to be steeper and cross the axis earlier. Eventually, when the object is very far away the ray will cross the axis at a distance f from the lens (which would be 10 cm in the earlier example). If we then move the object closer to the lens than the focal point, e.g. 7 cm (Figure 3.20, right), we can still use the deflection angle to find the image position, even though in this case the image distance is on the same side of the lens as the object: the image is virtual!

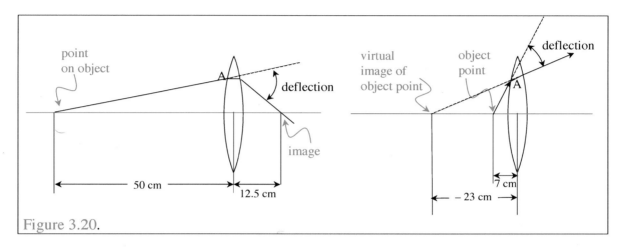

Figure 3.20.

3. Object Distance Smaller than the Focal Length:
Using the same lens, move the object to a distance 7 cm from the lens, as shown in Figure 3.20, right.

$$\frac{1}{i} = \frac{1}{f} - \frac{1}{o} = \frac{1}{10} - \frac{1}{7} = \frac{7}{70} - \frac{10}{70} = \frac{-3}{70}$$

image distance i = –70/3 = –23 cm

The minus sign indicates that the image is on the same side of the lens as the object, rather than on the opposite side as in examples 1 and 2. This *negative image distance i* indicates that the image is *virtual*. The size of the virtual image is 3.3 times the object size, since $M = \frac{i}{o} = \frac{23}{7} = 3.3$. For a converging lens, the image is virtual whenever the object distance is less than the focal length of the lens.

4. Now we are going to use a *diverging lens* with a focal length f = -10 cm. This can have different curvatures, as shown in Figure 3.21, but it must be thinner in the middle than at the edges.

We must use f = -10 cm (negative focal length) in the lens formula.

$$\frac{1}{i} = \frac{1}{-10} - \frac{1}{o} = \frac{1}{-10} - \frac{1}{o}$$

Note that the image distance i will always be negative (and the image will be virtual), no matter where we place the object—that is, for any value of the object distance o. For example, if o = 10 cm

$$\frac{1}{i} = \frac{-1}{10} - \frac{1}{10} = \frac{-2}{10} = -\frac{1}{5}; \text{ therefore } i = \text{-5 cm}$$

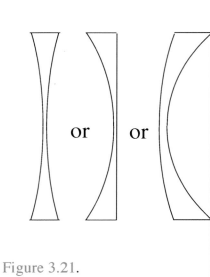

or   or

Figure 3.21.

The *virtual* image is 5 cm from lens, on the *same side of the lens* as the object. The image will be half the size of the object, because $M = \frac{i}{o} = \frac{5}{10} = \frac{1}{2}$.

Remember to neglect any minus signs when calculating the magnification.

Note on Magnification

As previously mentioned, the magnification is always lower than 1 for diverging lenses. In other words, diverging lenses are always demagnifying lenses. On the other hand, the magnification can vary dramatically for converging lenses, which can be magnifying or demagnifying depending on where the object is positioned with respect to the lens. See Figure 3.22 for various object positions. In these diagrams, notice that one object (the man with the umbrella) moves, while another object (the tree) is stationary. Compare the relationship between the two objects on each side of the lens for each diagram.

Figure 3.22.

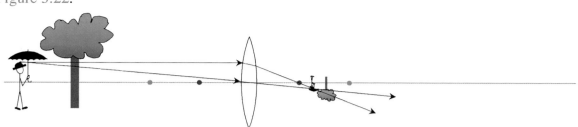

When the object (man with umbrella) is further than the focal length (blue dots), the image is real, upside down, and smaller than the object.

When the object is at twice the focal length (magenta dots) from the lens, the magnification is 1.

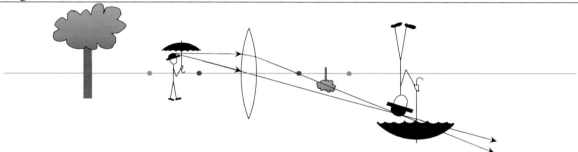

When the object is between the focal length and twice the focal length (between magenta and blue dots), the magnification is greater than 1.

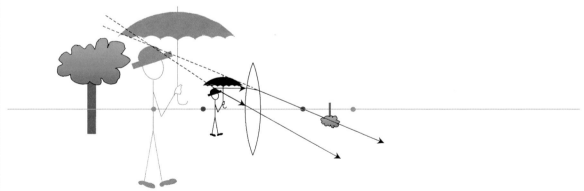

When the object is closer to the lens than the focal length, the image is virtual (gray man), upright, and the magnification is greater than 1. (Diagrams adapted from Eugene Hecht, *Optics*, Addison-Wesley Publishing Company, Reading MA).

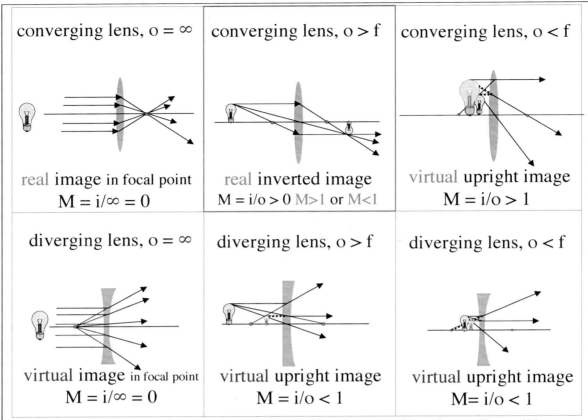

converging lens, o = ∞	converging lens, o > f	converging lens, o < f
real image in focal point $M = i/\infty = 0$	real inverted image $M = i/o > 0$ M>1 or M<1	virtual upright image $M = i/o > 1$
diverging lens, o = ∞	diverging lens, o > f	diverging lens, o < f
virtual image in focal point $M = i/\infty = 0$	virtual upright image $M = i/o < 1$	virtual upright image $M = i/o < 1$

Figure 3.23. Summary of the images and magnifications resulting from all possible cases for converging and diverging lenses: object at infinity, object farther than the focal length, and object closer than the focal length.

LENS ABERRATIONS

The formulas and ray tracing method we have discussed so far work accurately as long as the lenses are thin and, more importantly, as long as we are in the *paraxial approximation*, meaning that we only consider *rays of light close to the optical axis*. Selection of only paraxial rays can be done either using a lens with a very small diameter or using a larger lens with an iris aperture that limits the lens diameter actually used. Decreasing the effective diameter—or closing the aperture—limits the brightness of the image, that is, the *luminosity*. Consider an object and a lens. Rays from the object hit all points on the surface of the lens, not just the points hit by the three easy rays we considered before. Each and every portion of the lens is producing the image of every point of the object. You may therefore use a small lens to take a picture of a large tree or you may cover a portion of the lens and still obtain the same image! If, for example, you cover half of the lens, the other half (which now has a D shape), you will obtain the exact same image of the object, with same magnification and position, as if the entire lens was uncovered. This image, however, will have half the intensity. The reason for using a larger lens is to increase the lens's luminosity. That is why the best telescopes—those that can see the faintest stars—have very large diameters. The same applies to camera lenses: the best cameras have lenses with the largest diameters, allowing them to take photographs in dim light conditions by opening the aperture all the way and using the entire diameter of the lens.

Departing from the paraxial condition, for instance when taking a photograph with the aperture wide open, introduces problems in image clarity and shape. These problems are called *lens aberrations*. There are different kinds of aberrations. *Chromatic aberrations* arise from the fact that the index of refraction n varies for different wavelengths of light (colors) *Monochromatic aberrations,* on the other hand, occur even if a single wavelength of light is used. Monochromatic aberrations include: *spherical aberration*, coma, astigmatism, field curvature, and *distortion*. Here we will only focus on the types of aberrations that mostly affect photography.

Chromatic aberrations:
As described in Chapter 2, the index of refraction of glass is a little different for each different wavelength of light, a phenomenon known as *dispersion*. Rays of different colors are separated by the lens, as blue light rays are deflected more than red light rays. Consequently, the focal length of a simple lens for blue light is different than the focal length of the same lens for red light. The difference is not very large but it is enough to cause trouble. For example, you cannot capture both red and blue polka dots of a woman's skirt with a good, sharp focus at the same time if you are using a camera with a simple lens. If you focus on the red dots, the blue dots will become a little fuzzy and vice versa. If you can see the image of the red dots as perfectly in focus, the image of the blue dots will be in focus at about 0.2 mm in front of the film. By the time the rays get to the film they will have already re-diverged a little, producing a slightly blurry image of the blue dots. The blur is worse if the lens aperture is large because the rays diverge more over the 0.2 mm distance. This is because rays of light farther from the axis of the lens must be deflected more (at a sharper angle) to be brought into focus. The more rays of light are deflected by the lens, the greater the effect of chromatic aberration. If you don't need a large lens aperture, however (for example, if you only want to take pictures outside on sunny days), there is a simple way around this problem—just use a small lens aperture.

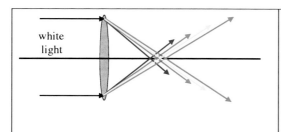

Figure 3.24. Blue light is deflected by the lens more than red light, so the focal point for blue is closer to the lens than the focal point for red. Blue and red objects cannot be in focus at the same time. Closing an iris aperture, i.e. selecting the central or *paraxial rays* only, reduces chromatic aberration.

Most cameras have *achromatic lenses* that produce sharper pictures and that can be used in dimmer light. Achromatic lenses are built by combining a strongly converging lens, made of low-dispersion glass, with a weakly diverging lens of high-dispersion glass. With this combination, the dispersion cancels out while net focusing remains. You might wonder why lens makers don't simply use a type of glass with no dispersion. There is no such glass or plastic! Another common question: Why not cancel the dispersion by making half of the lens with glass of the opposite characteristic? Again, this is impossible since all transparent materials deflect blue more than red.

Spherical aberration:

This type of aberration is color-independent and arises from the fact that rays hitting the *outer diameters* of a lens are *deflected more* than paraxial rays, the rays hitting the lens inner diameters. This is true for both converging and diverging lenses. The net result does cancel if you use a combination of diverging and converging lenses, as shown in Figure 3.25.

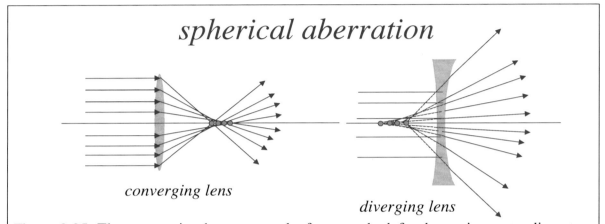

Figure 3.25. The converging lens moves the focus to the left when using outer diameters of the lens. The diverging lens moves the focus to the right when using the outer diameters. Since these two lenses compensate for each other, in camera lenses a combination of diverging and converging lenses is used to minimize spherical aberration. Unlike those in Figure 3.9, these must have different curvatures and focal lengths.

Another solution to eliminate spherical aberration is to use a small aperture, but again, this can only be done in conditions of bright light.

Chromatic and spherical aberration, as well as *coma* and *astigmatism*, deteriorate the image clarity. *Field curvature* and *distortion*, on the other hand, deteriorate the image shape. In other words, they deform the image.

Distortion: It is not enough that a picture is in sharp focus for all colors and for all rays passing through the lens. Suppose you take a picture of a rectangular building with a rectangular grid of windows. These should produce an image on the film of rectangles, not slightly curved lines. However, there are two types of common distortions that cause such curved images: the pin-cushion distortion and the barrel distortion, shown in Figure 3.26.

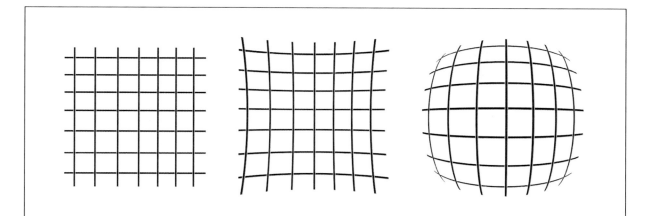

Figure 3.26. On the left is the object, at the center its pin-cushion distorted image, on the right its barrel distorted image.

These distortions arise from the fact that the image *magnification varies with the image radius*. With pin-cushion distortion, the center of the image has lower magnification than the edges, while with barrel distortion, the center has higher magnification than the edges.

The design of a good quality camera lens is complicated (and therefore the lens is very expensive) because it has to compensate for all the possible aberrations for all possible object distances, including objects near the center of the field of view and near the edges. The solution to the problem lies in making *compound lenses*. This is done by stacking different lenses, each lens curved and spaced from the next lens in the proper, particular way. Computer ray tracing programs are used to do trial and error simulations of compound lenses to find the combination of lenses with the best performances, thereby minimizing distortions and spherical aberration. Only then are the lenses physically mounted in stacks. For most purposes, it is not worthwhile to buy the absolute best lens available since the imperfections for even an average quality lens are only very slight. Even cheap lenses can take pictures of good quality, as long as the lens aperture is kept small. However, if your interest is to take high quality or professional photos when you really can't use a tripod or a flash, such as early morning wild life photography, you must use larger lenses. For example a lens with a very large diameter (e.g. 300 mm) that increases luminosity and allows for a fast exposure (which you need with no tripod and no flash). Such lenses are always compound, and you may have to spend more than $10,000 on a single lens! Not only is it expensive, but it is heavy, sometimes weighing 20 pounds! Imagine carrying that around the woods on long hikes. The life of a photographer is not as easy as one may speculate!

Coated lenses: All high-quality camera lenses have coatings to reduce the amount of light intensity that is lost to reflection from the lens surfaces. These are *non-reflective coatings*. Some lens coatings may have a noticeable magenta tinge. This is because a small amount of blue and red light is reflected and makes the coating appear magenta. However, the best quality coatings are perfectly colorless, and they do not allow for any light to be reflected. Let us see in detail why such non-reflective coatings are necessary to produce a good photograph.

Not all the light incident on the surface of the glass lens penetrates the lens and gets refracted. Part of the light is reflected back. The loss of light intensity on the camera film from this reflection is quite serious because, as more than one lens is usually used to avoid aberrations, the light goes through many surfaces and the amount of light intensity lost at each surface adds up.

Example:

If a compound lens is composed of 6 lenses, there are actually 12 surfaces.
Even if *only 8%* of the light intensity is lost to reflection on each individual surface,

$$0.92^{12} = 0.367$$

after traveling through all 6 lenses, only 36.7% of the original light reaches the film!

Without a non-reflective coating, you would really be better off using a simple lens and avoiding aberrations instead, and simply use a small aperture!

This is why, especially for compound lenses, a non-reflective coating is always used. The coating consists of a thin film that has just the right index of refraction and just the right thickness so that reflected light from the front and back surface of the coating cancel by destructive interference. The light cannot be reflected and has to go through the lens.

Chapter 4
THE EYE

The lenses discussed in Chapter 3 are useful in understanding how the eye works, its defects and the corrections that appropriate lenses can provide. The text in this chapter was adapted from Eugene Hecht,

Optics, Addison-Wesley Publishing Company, Reading MA, which we recommend for a more in-depth description.

The human eye is a complex organ composed of several parts, arranged as shown in Figure 4.1.

Rays of light cross a system of two lenses and then converge, in focus, on the retina to form *a real inverted image*. The image has to be real, that is, actual rays must converge on the retina. If the image is not real and not formed by actual intersecting rays, the cone and rod cells in the retina will not be able to detect the image, convert it into a train of pulses, and transmit it to the brain via the optic nerve. The left and right eyes produce different upside down images. The two images are then blended together and inverted again by the brain. The two images seen by the two eyes are identical to each other when the objects are very far away but their difference increases as object become closer. We do not realize that our two eyes see two different images until we close them alternately and observe the right and the left image separately. The difference in the images is important because it makes it possible for our brain to observe the three-dimensionality of nearby objects, and to provide an estimate of their distance.

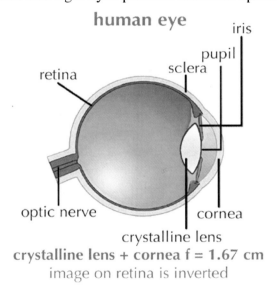

human eye

retina — iris — pupil — sclera — optic nerve — cornea — crystalline lens

crystalline lens + cornea f = 1.67 cm
image on retina is inverted

Figure 4.1.

The eye is contained within a resistant yet flexible eyeball, *the sclera*, which contains jelly-like fluids. Except for the front area, called *the cornea*, the sclera is white and opaque. The cornea is the curved surface bulging out from the eyeball. The cornea is slightly flattened at the center to reduce spherical aberration. It is a hard, transparent, convex layer of cells, forming a converging lens. It is the stronger of the two lenses in the eye, so it has a lower focal length and a higher curvature than the other lens. Most of the deflection of light rays by the cornea takes place at the air-cornea interface because the refraction index of the cornea ($n_{cornea} = 1.376$)

is much greater than that of air (n_{air}=1). It is similar to that of water (n_{water}=1.33), which is why we can't see well underwater: rays are not deflected much at the water-cornea interface and reach the retina out of focus. After the cornea, is *the iris*, which is immersed in a watery fluid and serves as a lens aperture. Just like the aperture stop in a camera, the iris that controls the amount of light entering through *the pupil*, or the hole allowing light to enter the eye. It is made of circular and radial muscles so it can expand and contract. Its diameter varies between 2 mm in bright light and 8 mm in darkness.

Immediately behind the pupil of the iris is *the crystalline lens*. This lens has the shape of a small lentil bean or an M&M candy, with a diameter of 9 mm and a thickness at the center of about 4 mm. The crystalline lens has a very complex structure, with 22,000 concentric layers (as in an onion), with indices of refraction that vary slightly from one layer to the next. At the dense, inner core, n = 1.406, while at the surface n = 1.386. A man-made lens with these characteristics is called a GRIN lens, from GRadient INdex.

The crystalline lens provides fine-focusing through changes in its shape. Behind the crystalline lens is a jelly-like fluid called *vitreous humor*, which differs from the fluid in front of the crystalline. The vitreous humor contains microscopic particles, which are dead cell debris, floating around within it. You can see these floaters by looking at the sky on a bright day: they appear as transparent, amoeba-like objects floating across the field of view.

At the back inner surface of the eyeball is *the retina*. This is a thin, semitransparent layer of light receptor cells. The human eye has two kinds of light sensitive cells: *rod cells* and *cone cells*. The rods provide low-resolution vision and function well in dim light, while the cones provide high-resolution color vision but only function well in bright light. There is a small depression of about 2.5-3 mm in diameter at the center of the retina that is known as the yellow spot, or the *macula*. This spot contains only cone cells and no rod cells. At the center of the macula is *the fovea*, with a diameter of 0.3 mm. To give you some perspective, keep in mind that the image of a full Moon on the retina is 0.2 mm. In the fovea the cones are thinner (with a diameter 1.5-3 μm) and more densely packed than anywhere else in the retina. Since *the fovea provides the highest resolution and most detailed visual information* at the center of our field of view, we continuously move our eyeballs so that light coming from the object of primary interest always falls on the foveas.

Each cone cell has its own connection with an individual nerve fiber, while approximately 100 rod cells connect to a single nerve fiber. All nerve fibers come together in the *optic nerve*, which brings the signal from the the retina to the brain. The optic nerve connects with all the nerve fibers converging from all over the retina. In the point in which the optic nerve is attached to the retina, there is a region in which we do not see. This is called *the blind spot*.

X   **1**   **2**

Figure 4.2.

NOTE: this may need re-sizing.

Look at the figure above, stand at a distance of about 10 inches, close your left eye, and focus your right eye on the X. You will only be able to see the number 1 because the 2 falls on the blind spot. If you now move closer, the 2 will reappear and the number 1 will vanish. You can test this on your left eye as well, but you have to rotate the book upside down.

ACCOMMODATION

As mentioned above, the crystalline lens provides fine-focusing of the image from the cornea and the pupil. The focal distance of the system of the two lenses in the eye, the cornea and the crystalline together, is 1.67 cm. If an object is at infinity, the image is exactly in the focal point, which is on the retina. If the object comes closer, the image will be formed in focus *after* the retina and the image on the retina will be fuzzy. The crystalline lens, which is flexible, can compensate for this effect. It becomes thicker in the middle, becoming a stronger lens with a smaller radius of curvature and a shorter focal distance. After the crystalline lens thickens, the images will be perfectly in focus on the retina. This process of thickening is called *accommodation*.

Figure 4.3.

accommodation of the eye

near object: the crystalline lens thickens   far away object: the crystalline lens is relaxed

but there is a limit: the "near point"
this is the shortest distance of distinct vision

As the object moves closer to the eye, the crystalline lens accommodates more and more until it reaches a maximum point beyond which it cannot thicken anymore. That maximum crystalline thickness defines *the near point* of the eye. This is the shortest distance of distinct vision. You can easily find your own near point. With one hand cover one eye, extend your other arm, and look at your finger. Now, bend your arm and slowly bring that finger closer to the eye, keeping it in focus. As you do this, your crystalline lens is thickening! As the finger approaches the eye, it will get closer and closer to the near point. When it passes the near point, you will not be able to focus on the finger anymore.

The near point—the shortest distance at which you can focus—varies from person to person, and it also changes for the same person with age. As a person gets older, the crystalline lens loses elasticity and the near point is at greater distance away from the eye. With a normal eye, the near point is 7 cm for a teenager, 12 cm for a young adult, 28-40 cm for a middle-aged person, and about 100 cm for a 60 year-old. That is why the normal eye after a certain age needs reading glasses. These move the near point closer. As you may have imagined already, this correction is done using glasses with converging lenses. The myopic eye may never need reading glasses.

All mammals accommodate their vision in this way. Fish, on the other hand, can only move the lens in their eye away or towards the retina to adjust the focus, just as we do when using a camera. Some mollusks elongate or contract the eyeball depth, and birds of prey accommodate by changing the curvature of the cornea.

EYEGALSSES

Macfarlane, an anthropologist, and Martin, a historian of glass instruments, both from Cambridge, United Kingdom, speculate that the development of glass was fundamental for science itself: lenses are the major components of telescopes for looking at the stars and of microscopes for looking at cells. Imagine life without glass, before the first synthesis of transparent acrylics and plexiglass. There would have been no windows to keep the warm air inside while letting in sunlight, therefore the time for reading and studying would have been far shorter, slowing down progress considerably. There would have been no bottles to foster the development of alchemy and chemistry. Ancient cultures that drank tea instead of wine, such as the Chinese and the Japanese, never had a good reason to develop glass blowing skills because ceramic pottery had far superior characteristics to hold hot liquids without cracking. Macfarlane and Martin infer that the stagnation of those cultures in and after the Middle Ages was due to the absence of glass. However, these are only speculations and there is of course no proof of how the development of glass technology would have affected those cultures.

Eyeglasses were invented first in Florence, Italy, in 1299. These were convex, converging lenses only. Glasses using diverging lenses did not appear until about 150 years later. Negative, diverging lenses where actually first produced in 1267, but they were not used in glasses for about 200 years. Interestingly, it was considered improper in Europe to use eyeglasses in public until the 1800s and, therefore, we have very few paintings of people wearing glasses.

A *normal eye* is an eye capable of focusing parallel rays on the retina while in relaxed condition.

The *dioptric power* [DP = 1/f (m)] of the unaccommodated normal eye is 60 D (1 diopter = 1 D = 1 f(m)$^{-1}$), of which 43 D is from the cornea and 17 D is from the crystalline lens. 60 D corresponds to a combined focal length of the two lenses of 1.67 cm (1/60 × 100 cm).

The *far point* is the maximum distance of a point whose image lies on the retina, or the maximum distance of distinct vision. For a normal eye, the farthest point focused on the retina, the far point, is located at infinity. Infinity, for all practical purposes, is anywhere beyond about 5 m from the eye's lens. When the focal point of the eye's system of two lenses does not lie on the retina, the eye is nearsighted, farsighted, or astigmatic. This may be a result of abnormal changes in the lenses (refractive index or curvature) or the dimensions of the eyeball, which alter the distance between the lens system and the retina. The second problem, abnormal dimensions, is the most common cause of imperfect vision.

NEARSIGHTED EYE

A deep eyeball causes the eye to be *nearsighted, or myopic*. This can be corrected using a negative lens. In a nearsighted eye, parallel rays converge before the retina and have diverged again by the time they reach the retina. Therefore, far away objects appear blurry and larger than they should. The far point is closer than 5 m, for instance at 2 m. To correct for this problem, we have to add a lens to the optical system. Only when the three lenses are combined (the eyeglass lens, the cornea, and the crystalline lens) can we produce an image of parallel rays in focus on the retina. Since the nearsighted eye cannot focus on far away objects, we add a *diverging lens* that forms a virtual image of distant objects closer to the eye. For objects at infinity, the lens creates a virtual image located at the far point (2 m). No accommodation by the crystalline lens is required. If the object is closer than infinity, the virtual image is closer than the far point, say 1 m away in the above example. This image is then viewed by the cornea and crystalline lenses, which cast a real, focused image on the retina, but only after accommodation. Another result of using these diverging lenses is that they move the near point a little further. This is why nearsighted people prefer to take their spectacles off when performing precision tasks, such as threading needles or reading small print. Without the diverging lens, they can then bring the object closer to the eye, and increase the magnification.

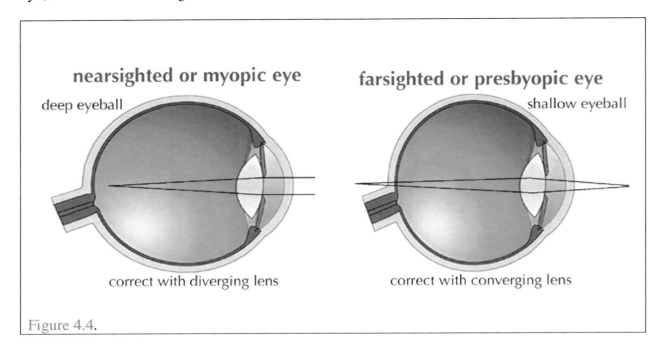

nearsighted or myopic eye   farsighted or presbyopic eye

deep eyeball   shallow eyeball

correct with diverging lens   correct with converging lens

Figure 4.4.

FARSIGHTED EYE

A shallow eyeball causes the eye to be *farsighted,* or *hypermetropic*. Objects at infinity are seen clearly, but nearby objects are blurred because the rays converge after the retina. Recall that the crystalline lens accommodates only up to a point, reaching its limit at the near point. The near point is further than norm for a farsighted eye. This defect can be corrected with a *converging lens*, which makes the rays converge earlier, on the retina. Such lenses effectively move a close object further away, beyond the near point, where the eye can see clearly. The lens will form a distant, virtual image that can be clearly seen by the cornea-crystalline lens.

Recall that diverging lenses are always demagnifying, while converging lenses with nearby objects are always magnifying. You can, therefore, always tell, when looking at a person in front of you wearing eyeglasses, which kind of lenses they are using. In this case the person's eyes are the objects, then the lenses are their eyeglasses and the image is the image of their eyes, which you can see standing in front of them. If their eyes appear smaller through their glasses the lenses are diverging, therefore you know the person is nearsighted. If their eyes are magnified by their glasses you know the lenses are converging and they must be farsighted.

For the same reason, nearsighted people always see the world a lot smaller when they wear their glasses. This is not only due to the fact that sharp objects are smaller than blurry ones. It is also due to the demagnifying effect of all diverging lenses.

ASTIGMATIC EYE

Astigmatism is the most common eye defect. Almost everyone has some degree of astigmatism, but in most cases this is so slight that people don't even know they have it. It arises from a non-spherical curvature of the cornea. In particular, the curvature along different meridians differs. Most frequently the curvature of the cornea along a meridian is greater than the curvature in the plane perpendicular to that meridian. The result is a blurred and/or distorted image on the retina. This defect can often be corrected using *cylindrical lenses*, first introduced in 1825 but only widely spread after 1862.

Chapter 5
PHOTOGRAPHY

THE CAMERA

The oldest, preserved, black and white photograph was taken by Joseph Nicéphore Niépce in 1826. It was produced by a dark box, a lens and light from his window view projected onto a pewter plate, coated with bitumen, a petroleum derivative. After a day-long exposure, the plate was washed with a mixture of oil of lavender and white petroleum which dissolved away the parts of the bitumen which had not been hardened by light. The result was the permanent direct positive photograph on pewter, rendering a view of the outbuildings, courtyard, trees and landscape as seen from that window, which can still be see today at the University of Texas at Austin. Soon after this first photograph was taken, Daguerre (from whom the name "daguerreotype" for photographic prints on glass is derived) and William Henry Fox Talbot invented the photographic process in which a *negative* was exposed in the camera, developed in a dark room and then printed on sensitive glass or paper. Their cameras were conceptually similar to modern ones, and are the easiest of the optical instruments to understand. The key parts that form any camera, be it the simplest ancient version or a very sophisticated modern one, are shown in the schematic diagram of Figure 5.1. These are:

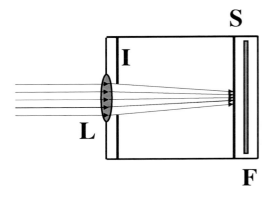

Figure 5.1.

- *The lens,* **L,** produces a *real image* of the object that one wishes to capture on the film. It is usually mounted on a barrel cap that can moved in and out, allowing the photographer to adjust the distance between the film and the lens in order to focus the image.
- *The shutter*, **S,** is normally closed but opens for a small interval of time to allow light from the object to enter the camera. The shutter may be placed near the lens ("lens shutter") or near the film ("focal plane shutter").
- *The iris aperture,* **I**, also called *f-stop,* is an adjustable diameter aperture (hole or orifice) located near the lens. This controls the amount of light that enters the camera.
- *The film,* **F,** consists of a base, a polymer sheet or glass plate, that is coated with a light sensitive *emulsion.* This emulsion contains grains of silver salts (silver halides, such as AgBr) in a gelatin.
- When light strikes the emulsion, the emulsion grains are modified. Chemical treatment of the emulsion, including *developing* and *fixing*, removes the non-

modified emulsion grains and transforms the exposed grains into metallic silver, which is black.

Let us now look in more detail at the operation of the various parts of the camera.

FOCUSING THE CAMERA

Focusing the camera involves adjusting the distance between lens and film so that the *real image* of the object being photographed is located on the film plane, not in front or behind it. Why is that important? Consider a point object being photographed. If the camera is not focused properly, the light rays will not converge to a point on the film. Rather, they will produced a fuzzy spot of a larger size, and the picture will be blurred.

For an example, consider a simple *35 mm camera*. This label refers to the size of the film, which is 35 mm wide and 24 mm high. The focal length of the lens is most commonly f = 50 mm. If you wish to photograph a far away mountain, the object distance is very, very large. When the object distance is more than 100 times greater than the focal length of the lens (5 m in this case), photographers refer to the object as being at infinity. They use the standard mathematical symbol for infinity to represent this distance: ∞.

When the object is very far away, we write the lens formula as follows:

$$\frac{1}{f} = \frac{1}{i} + \frac{1}{o} \qquad o = \infty \qquad \text{then} \qquad \frac{1}{o} = \frac{1}{\infty} = 0 \qquad \text{and} \qquad i = f$$

In other words, if o is very large, $\dfrac{1}{o}$ is very small—so small that it can be neglected! The image distance is then equal to the focal distance. Therefore, when using a lens with a focal distance f = 50 mm, both the film and the image must be at 50 mm from the lens in order to take a photograph of a mountain.

Suppose now that you want to take a close-up picture of a flower about 20" away.
20"≈ 0.5 m   o = 0.5 m = 500 mm
                        f = 0.05 m = 50 mm

Using the lens formula again:

$$\frac{1}{f} = \frac{1}{i} + \frac{1}{o} \qquad \frac{1}{i} = \frac{1}{f} - \frac{1}{o} = \frac{1}{50} - \frac{1}{500} = \frac{10-1}{500} = \frac{9}{500} = 0.018 \quad \text{and} \quad i = 55.55 \approx 56 \text{ mm}$$

The distance between the lens and the film must be 56 mm. As you can see, a very small motion of the lens, in this case 56 - 50 mm = 6 mm (or about 1/4 inch), is all that is needed to change focus from infinity to 0.5 m (two feet). Thanks to the fact that only a small variation in image distance is necessary to focus objects at any distance, cameras can be small and portable!

The change in image distance with focusing is small, but by no means negligible. Light rays diverge a lot in a small distance, therefore accurate focusing must be done, otherwise the image will be blurry. The Shutter is placed either near the lens or near the film. The *focal plane shutter*, is a double curtain mechanism that lies directly in front of the film—i.e. as close as possible to the focal plane. Initially, both curtains cover the film (frame 2 in Figure 5.2). When you press the button to take a photograph, the first curtain starts to move across the whole film width, exposing the film to the light entering from the lens (frames 2-7). After a certain time (which the photographer sets as the *exposure time*), the second curtain starts to move in the same direction as the first curtain, covering the film again and cutting off light exposure (frames 8-12). For short exposures, the two curtains move very closely together, keeping a constant distance between them. Each point in the film is exposed for the exact same amount of time (6 time-steps in Figure 5.2). In real cameras, the curtains move continuously, not in steps as in the schematic representation of Figure 5.2. The sequence reported is only a simplification to aid your understanding.

Figure 5.2.

All cameras have a set of numbers that indicate the *exposure time*, or the time that each point on the film is exposed to light as the shutter curtains move across the front of the film. The exposure times are usually: 1000, 500, 250, 125, 60, 30, 15, 8, 4, 2, 1, B, T. This actually means $1/1000^{th}$ of a second, $1/500^{th}$, $1/250^{th}$, etc…. down to 2, meaning 1/2 of a second and 1 meaning simply 1 second. These are the times a photographer has at his disposal. Notice how they form a sequence varying by factors of 2, approximately.

Exposure time in seconds	1	1/2	1/4	1/8	1/15	1/30	1/60	1/125	1/250	1/500	1/1000
Geometric progression	1	1/2	1/4	1/8	1/16	1/32	1/64	1/128	1/256	1/512	1/1024

Exposure times vary by factors of 2. For each step toward the left, the exposure time doubles, and to the right it is cut in half. This is not perfectly accurate, as you can see by comparison with the geometric progression in this table, but it is very close. We will soon see how this is very useful when selecting the appropriate settings in order to capture the desired photograph.

When a camera is set on B, which stands for bulb, the shutter is opened when the button is pressed and closes again when the button is released. When the camera is set on T, which stands for time, the shutter is opened by pressing the button and closes only by pressing it again.

CHOOSING THE EXPOSURE TIME

If the object to photograph is moving, it may appear blurred depending on how much it moved while the shutter was open. The distance that an object moved is equivalent to the speed × time. In other words, d = v × t. A track star running a four-minute mile travels at a speed of 7 m/sec. If you set your camera's exposure time to 1/100 sec, the runner will have moved about 3" during the exposure time and the photograph will look quite blurry. This reveals that the exposure time of 1/100 sec is too long (or too slow) to take a good action shot of the runner.

$$7 \text{ m/sec} \times 1/100 \text{ sec} = 0.07 \text{ m} = 7 \text{ cm}$$

A person walking at 1 m/sec moves only 1 cm in 1/100 sec. It is therefore acceptable to take photographs using an exposure time of 1/100 sec with people moving at this speed.

Even when the object is stationary, if the camera is held by hand, the maximum exposure time that can be used is about 1/60 sec. This is because of the natural and inevitable movements of the human body. For any exposure time longer than 1/60 sec, as necessary for taking photographs at dusk, special attention is required to not move the camera. If the camera moves, the image of the stationary object moves on the film and the image will appear blurred. At 1/60 sec there is not a blurring problem for any hand-held camera. Even at 1/30 sec, if you hold your breath and lean against the wall you may be able to take a good picture. However, for longer exposure times, such as 1/15 sec or 1/2 sec, the camera must be held by a tripod or set on a table or car top.

CHOOSING THE APERTURE

The iris aperture is a hole, or orifice, located immediately after the lens. It is made of many thin metal blades that can be pivoted along the rim in order to adjust the diameter of the opening. Cameras normally do not show the actual diameter of the aperture. What is shown, printed on the lens mounting, are the various *f numbers* that are mathematically represented by the symbol *f/*. The diameter of the iris aperture is calculated by:

$$f/ = \frac{f}{Diameter}$$

where f is the focal length of the lens and Diameter is the diameter of the aperture—the effective diameter of the lens portion being used. For example, if a camera has a lens with 50 mm focal length and the aperture is set at an f number of 4 (this is written *f/4, not f/=4*), then the diameter is actually 50 mm / 4 = 12.5 mm.

The sequence of *f numbers,* or *f/,* that normally appears on the aperture adjustment control is:

f/	1	1.4	2	2.8	4	5.6	8	11	16	22	32	45

But what happens when you change the f number by two steps, from f/4 to f/8? That is, from 4 to 5.6 (one step) and again from 5.6 to 8 (another step). A photographer calls this change "two stops" because *aperture stop* is the colloquial, commonly used name for the iris aperture. The diameter of the aperture is changed in this example from 50/4 = 12.5 mm to 50/8 = 6.25 mm.

Through comparing these two diameters, it is apparent that changing the aperture by 2 steps changes the diameter by a factor 2. The *area* of the aperture changes by a factor of 4 (because the area of a circle is proportional to the radius squared).

Changing the aperture by 1 step changes the diameter by a factor of 1.4:

$$\frac{2.8}{2} = 1.4; \qquad \frac{4}{2.8} = 1.4; \qquad \frac{5.6}{4} = 1.4; \qquad etc.$$

One step changes the area by a factor of 2:

$$1.4 \times 1.4 = 2.$$

Each single step in f/ changes the lens area two-fold, and therefore also changes the amount of light on the film by a factor of 2.

Besides its primary function of controlling the amount of light entering the camera, the value of the f number also affects another aspect of the picture: *the depth of field*, defined in Figure 5.3.

DEPTH OF FIELD

Suppose you are taking a photograph of an object that is 4 m away from the camera. If you want to take a good picture, you will naturally want to adjust the lens so that the object is in perfect focus. Objects farther away and in the background, say 5, 7, 8 or more meters from the main object, will be out of focus. These background objects will become increasingly more and more blurred the farther they get from the main object you are focusing on. The same is true of objects closer by and in the foreground, say 3, 2, or 1 meters away

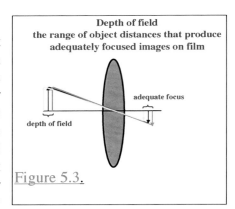

Depth of field
the range of object distances that produce adequately focused images on film

adequate focus

depth of field

Figure 5.3.

from the main object. Now, if the f number is large (indicating a small diameter), you will find that objects at both 10 m and 3 m are in acceptable focus—both are not too blurred. Objects farther than 10 m or closer than 3 m, however, are blurred on the photograph. Therefore, we say that the *field* extends from 3 to 10 m and that the *depth of field* is 7 m.

Figure 5.4. Large f numbers give large depths of field, and conversely, small f numbers give small depths of field. The reason for this is easy to understand. Consider an object whose real image is not on the film but behind it, as the red man with the umbrella in this diagram: this object is *out of focus*. The green man is instead focused in front of the film, and his image is also blurry on the plane of the

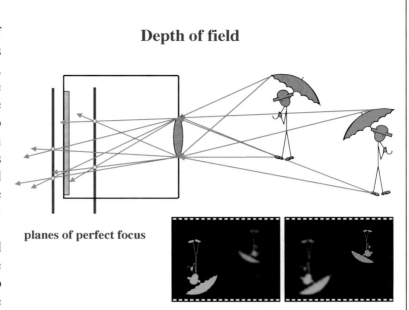

Depth of field

planes of perfect focus

film. If we focus on the green man, the red man will be completely out of focus, and vice versa. The two resulting negatives will appear as shown here with one man in focus or the other, but not both in the same photo. The images are upside down, as always in cameras, and the colors are inverted on the negatives shown here. Also, the far away green man appears smaller in the negative (magenta man).

Figure 5.5.

Depth of field

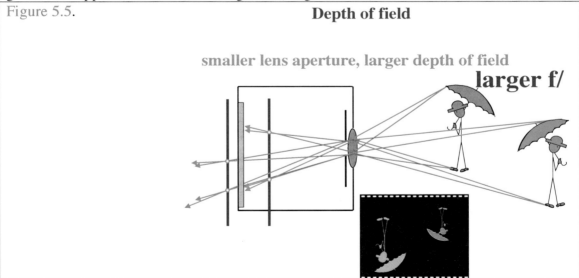

smaller lens aperture, larger depth of field

larger f/

As you can see, taking a photograph with the aperture wide open (Figure 5.4), both objects that are nearby and further away are in focus on their planes of perfect focus. Notice how the green rays from the far object (green man) cross on the first plane of perfect focus. By the time they reach the film, however, they are diverging again and the image of the green man will come out as blurred on the film. The red rays from the close object (red man) also converge on their plane of perfect focus. However, since this plane is after the film, the image the film captures will also be bigger and blurred. When closing the aperture (Figure 5.5), *the planes of perfect focus do not move—they are still* in front and behind the film. The difference is that now, rays can only enter the lens at small angles from the optic axis. Traveling at such acute angles, both the blue and magenta rays diverge very little: the image of far and close objects is small enough on the film that the focus is acceptable! The negative in this case shows both men in focus.

large depth of field
(large f/)

small depth of field
(small f/)

Figure 5.6. These two photos were taken with different f numbers (and correspondingly different exposure times). The top image shows flowers that are well in focus in both the foreground and the background (nearby and far away flowers, respectively). In the bottom image, a smaller f number selected only flowers at a very specific distance to be perfectly in focus. Flowers closer and farther than that distance appear blurred.

Why f number?

There is a perfectly reasonable question you may ask: "I understand why you need an aperture to control the amount of light entering the camera, and I also understand that steps of 1.4 in diameter correspond to steps of a factor 2 in the area of the aperture, BUT why not simply label the aperture stop with the different aperture diameters instead of talking about the mysterious f number?"

The answer is the following: *the amount of light on the film that matters is not the total, but the amount of light per square millimeter on the film.* You may think of this as the density of light rays. The f number is useful to keep this density constant. Therefore the f number was not introduced to make the art of photography more difficult, but to simplify it!

Consider the following example to clarify this point. Imagine a piece of paper with some orchids printed on it that is photographed by two lenses with focal lengths of 35 mm and 70 mm (or simply by a single zooming lens, allowing you to adjust the focal length from 35 to 70 mm—this is a very common zoom).

Figure 5.7. The illuminated area on the film is 4 times larger for the photograph taken with the 70 mm lens. If I want the same illumination on the film as for the 35 mm lens, I must have 4 times more light coming in through the lens. Therefore, the diameter of the iris aperture behind the lens must be twice as large for the 70 mm lens as it is for the 35 mm lens. In other words, the f number must be the same! You may switch lenses or zoom in and, if you keep the same f number, you can keep the same exposure time.

All lenses with the same f/ transmit the same amount of light per unit area of the film. This is quite convenient when you are using a zoom and changing the focal length of the lens because you do not need to recalculate or re-measure the exposure time with an exposure-meter!

THE FILM

You can buy films of many types, made by many manufacturers. Let us start with black and white (B&W) film for simplicity's sake. Color film will be explained in more detail in Chapter 8. All B&W films are coated with an emulsion of silver-halide grains. Those grains that were illuminated turn into metallic silver when the film is developed in the dark room. The more light strikes them, the more of them turn into silver. That is why the film produces negative images. These negatives are then inverted again to produce positive images when printed. Films differ by several properties, listed below:

- *Spectral Sensitivity*
  Some films are sensitive to blue and green light only (called orthochromatic film), while others are sensitive to red light as well (called panchromatic). The advantage of ortho film is that it can be developed in the darkroom under a red "safety light" so you see the progress of the development. There are even special types of film that are made to be sensitive to the infrared or the ultraviolet wavelengths. Most black and white films that you can buy are panchromatic. Color films are obviously sensitive to all colors.

- *Contrast*
  Suppose that you take a picture of a person wearing a relatively dark shirt that causes the light coming from the person's face to be ten times stronger than the light coming

from the shirt. On the developed film will be a dark face (inverted!). If the image of the face on the film is ten times darker than that of the shirt, we say that the contrast is normal. If the face is more than ten times darker than the shirt, however, we say that the negative has high contrast, and if it is less than ten times darker, we say that the negative has low contrast.

- *Grain*
  The size of the original silver halide crystals and the method of development determine the final size of the silver specks that form the negative image on the film. This size is called the grain size, and corresponds to the resolution or number of megapixels in digital photographs.

- *Speed*
  Different types of film require different amounts of light to produce the same amount of darkening on the negative. A "fast" film requires a small amount of light to darken the negative, while a "slow" film requires a lot of light.

The speed of films is measured in *ISO* (International Standard Organization). Currently, ISO has been adopted in all countries and has replaced both A.S.A. units (American Standards Association) and DIN units (Deutsche Industrie Norm), The diagram below shows several ISO speeds that are commercially available.

Film speeds (ISO)

they *also* vary by factors of 2

| 50 | 100 | 200 | 400 | 800 | 1600 | 3200 |

slow film   fast film

Film speed, Sensitivity (equivalent to "more light" on film)

Figure 5.8.

Notice in Figure 5.8 that ISO rating is **inversely proportional** to the amount of light required to produce a certain standard black on the negative image. Doubling the ISO number by using a faster film allows you to reduce the amount of light you need to let into the camera by 1/2.

It may seem as if it is always a good idea to buy the fastest film, so that you can take photographs in dim conditions of light, but you have to be careful: the fastest films (e.g. ISO 1000) have a much larger grain size. Therefore, while the photograph printed on 4" x 6" paper might appear acceptable, if you tried to enlarge it, it would appear grainy and have a low resolution. Fast films also have lower contrast (lack of deep black) and lower tonal range, that is, fewer levels of gray between black and white, and fewer color tones in color photography.

Interestingly, the film speed is a useful concept non only when using film, but also for digital cameras. Most digital cameras, in fact, allow the photographer to select the ISO. Selecting higher ISO on a digital camera, just as in film photography, enables photographs to be taken in dim conditions of light. In this case the resolution is inferior, as for film, because of "binning", that is, the amount of light/unit area on the CCD is artificially increased by increasing the "binned" area. If 2×2 pixels are binned together, the amount of light with be 4 times greater, and so on. In other digital cameras the resolution is not decreased but the gain (amplification) is increased, therefore there is greater pixel noise.

Digital Photography

Digital photography has been embraced by artists, professional photographers and photojournalism, and of course, by the amateur "point-and-shoot" photographers. Digital photography is no longer in its infancy, and its resolution satisfies the needs of many professional photographers and artists. Large-format digital sensors with 30 or more megapixels are now available. Film-based photography, however, and in particular large-format cameras with film-sizes between 4'×5" and 8"×10", are still favored by landscape photographers. These cameras can reach an image quality considerably beyond that of the best digital cameras. The wonderful large silver-prints of nature photographs of the in the American south-west by Anselm Adam, printed from negatives and taken with a large-format camera, are unsurpassed.

Conventional films still have a considerable advantage when comparing the overall resolution, contrast, and tonal range of an image

Despite the current inferiority compared to film photography, in many instances the loss in image quality is so small and the digital cameras are so convenient to use that more and more people are abandoning film photography—the authors of this book included! The conveniences of digital photography are very appealing: you don't have to buy film, develop negatives print the photographs, store them, and in the end rarely have them handy when they are needed. It is so much easier to simply transfer the files to a computer, process, enhance and modify them as you please, and have them available at all times, stored in the small volume of a computer. We therefore look forward to further progress in the resolution, contrast and tonal range of digital photography.

The film photography industry is certainly trying very hard to compete with the digital world: films today are substantially better than the ones available only ten years ago. Several companies are currently experimenting with a new type of color film that could, in the near future, become ten times faster than current films without too much loss in color fidelity, resolution, or contrast. Continued research will lead to further improvements, and these may keep film photography competitive with digital alternatives. Film and digital photography now complement each other and they may continue to do so, allowing them to coexist and provide photographers, artists and other people with many choices for taking a picture.

Obviously, digital cameras do not require a film, and use instead a charge coupled device (CCD) to capture the image, an then transfer it to a magnetic memory card. The sensitivity of CCDs, however, is still measured in ISO and can be changed by the user, so all we say about photography here applies to both film and digital photography.

PUTTING IT ALL TOGETHER: TAKING A PHOTOGRAPH

As we have seen, there are three numbers to take into consideration to capture a good photograph: the exposure time, the f number and the film speed. The diagram in Figure 5.9 summarizes the values possible for each of these.

Figure 5.9.

For each one of these three parameters, the most relevant variable is the amount of light on the film, or illumination I, that results. The arrows in the charts above indicate the directions in which I increases as you change each variable. Again, whenever varying the *exposure time*, the *f number,* or the *film speed* by *one step, I* varies by a factor of *two.* This is quite convenient because it greatly simplifies all the calculations to be done before taking a photograph.

Once you have the film in the camera, the film speed cannot be changed. You are left with only the exposure time T and the f number f/ to alter, or to "play with" if you really like photography! The total amount of light I that falls on the film depends on:

$$I = \frac{L \times T}{(f/)^2}$$

where L is the amount of light reflected by the object being photographed and transmitted by the camera lens.

Remember, the f number of the lens is given by $f/ = \dfrac{f}{Diameter}$.

For an example, let us assume that we are taking pictures outside. Since we cannot change the amount of sunlight reflected from the object towards the lens, all we can play with is T and f/. A picture taken at 1/60 sec and f/8 will have the same exposure as one taken at 1/250 and f/4. Here is why:

$1/60 \rightarrow 1/125 \rightarrow 1/250$ is two steps *less* light on film

$8 \rightarrow 5.6 \rightarrow 4$ is two steps *more* light on film

Decreasing the exposure time from 1/60 to 1/250 *decreases* the amount of light by a factor of 4. To compensate, we open the aperture by two stops, *increasing* the lens area by a factor of 4.

Are these two pictures the same? The answer is no. The picture with f/8 (the larger f/) has a larger depth of field, and a tree 3 m away and a mountain 1000 m away would both be in focus. However, a child jumping off the tree would be blurred because of the long exposure time. In the second picture with the smaller f/ and shorter exposure time, the child would not be blurred but the mountain in the background would be.

Figure 5.10.

When taking a photograph, a larger f number requires a longer exposure time. This is because a large f/ gives less light on the film. This has the advantage of an increased depth of field, but with the longer exposure time it is not possible to photograph movement effectively. A short exposure time gives less light on the film but has the advantage of capturing movement much better. Figure 5.10 summarizes all parameters. *This diagram is fundamental for taking good photos and to solve any photography problem.*

A handheld exposure meter or the built-in exposure meter of a digital camera tells a photographer which settings will be best in the particular conditions of light that is being measured. The exposure meter gives the photographer two numbers (or sets of two numbers): an f/ and a corresponding exposure time. However, the photographer is not bound to use those settings only—he can use any other combination of f/ and T that produce the same, ideal amount of light on the film. If he increases the f/, he'll obtain a better depth of field, but he will also need to increase the exposure time. The same applies for capturing movement: if he shortens the exposure time to better catch motion, he must also lower the f/.

This is why commercial handheld exposure meters always give many combinations of f/ and T, depending on the meter one can set 1, 2 or 3 steps of overexposure or underexposure. Auto-exposure digital cameras (in auto mode) give only one particular combination of f/ and T. More sophisticated digital cameras give different combinations depending on the "scene" setting.

Furthermore, in modern cameras, one can select "exposure time priority" or "aperture priority". This means that one chooses T and the camera automatically adjusts f/, or vice versa.

This combination (or combinations) of f/ and T represent a "good setting" for a photograph, as measure by the exposure meter. Starting from the measured "good setting", you can calculate other settings, for example to give priority to short exposures, or to large f/, depending on the photograph you wish to take. Starting from the numbers suggested by the exposure meter, to adjust these settings we must move *to the right* on the sequence of f/ and *to the left* on the sequence of exposure times (Figure 5.10).

Let us look at an example:

An exposure meter tells us that the "good settings" for taking a photograph are f/22 and exposure time T = 1/30 second.

f/1 1.4 2 2.8 4 5.6 8 11 16 22 32 45

T = 1 1/2 1/4 1/8 1/15 1/30 1/60 1/125 1/250...

However, since we do not have a tripod with us, we actually need to choose a much shorter exposure time T to avoid blurring of the photograph, say 1/250 sec instead. (Remember, natural shaking of one's hand causes a photograph to be blurry if T is longer than 1/60 sec).

With this alteration, we have moved 3 steps to the right on the exposure time sequence (3 steps less light on the film, or 2 × 2 × 2 = 8 times less light). Therefore, we must compensate by moving 3 steps to the left on the f/ sequence (3 steps more light). The new photograph is now taken with f/8 and T = 1/250 sec. The disadvantage from this change is a smaller depth of field but an advantage is that we don't need a tripod and we can photograph action better.

Let's now consider a slightly more complicated case, in which all 3 parameters are changed.

The ambient light conditions determine that a good photograph is taken with 1600 ISO film, f/5.6, and T = 1/125 sec.

ISO 1600

f/1   1.4   2   2.8   4   **5.6**   8   11   16   22   32   45

T = 1   1/2   1/4   1/8   1/15   1/30   1/60   **1/125**   1/250 ...

However, we really want a better depth of field than that given by f/5.6. For the depth of field we want, we need to increase the f/ to f/8. However, we also ran out of film, and the only film in our bag is 200 ISO. This is a much slower film and it requires much more light than the 1600 ISO film that we were using before. This decreased sensitivity of the film is conceptually equivalent to "less light on the film," and therefore we have to compensate for this loss by increasing the amount of light from one or both of the other two parameters.

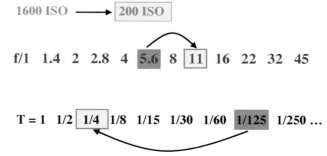

In total, from these changes we took 3 steps less light by changing film speed and 2 steps less light by changing the f/. Therefore, we must compensate by increasing the exposure time T by 5 steps (5 steps more light).

The new exposure time is T = 1/4 sec.

The disadvantage from this T is blurring when the camera is hand-held.

Chapter 6
COLOR AND COLOR VISION

COLOR

White light is a mixture of light of many different wavelengths. If you use a prism or a diffraction grating to break white light from the sun into its components, you see a sequence of colors that varies from red to violet.

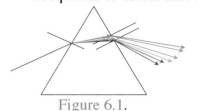

Figure 6.1.

The prism separates the different colors because *the index of refraction n is slightly different for each wavelength*, that is, for each color. This phenomenon is called *dispersion*. When white light illuminates a prism, the colors of the spectrum are separated and refracted at both the first and the second prism surfaces that light encounters. Refracted rays are deflected towards the normal on the first refraction and away from the normal on the second. If the prism is made of crown glass, the index of refraction for violet rays (n_{400nm}) =1.53, while for red rays, n_{700nm}=1.51. From Snell's law, we know that the greater the index of refraction, the more the rays are deflected. Therefore, violet rays are deflected more than red rays.

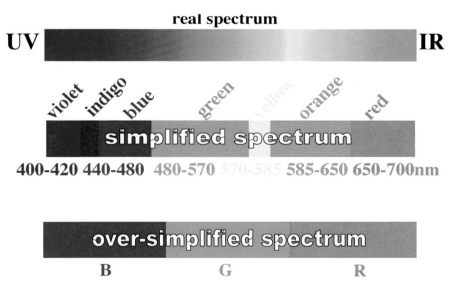

Figure 6.2.

The infinity of colors in the real spectrum (top panel of Figure 6.2) are called *spectral colors*. Of this infinity of continuously varying colors the human eye can distinguish up to 10 million colors. The second panel is a simplified version of the spectrum, with abrupt and **completely artificial** separations between colors. However, as a figure of speech we do identify quite a broad range of wavelengths as red, another as orange, and so on. Each range produces a specific sensation detected by our eyes, and the accurate wavelength ranges for each color are reported in the second panel of Figure 6.2. An even more extreme simplification is shown in the third panel of Figure 6.2. It is comprised of only 3 colors, R, G, and B, and is commonly used by computer monitors and TV sets. In these cases the three colors are added to produce a wide range of colors, as explained in Chapter 7. You may notice that many colors you know from everyday life are not present, even in the most complete sequence of spectral colors of the top panel. Purple, magenta, and mauve are one class of colors missing from the spectrum. Other missing colors are all the browns and earth colors, olive and hunter greens, and all the pastel colors such as pink, cyclamen, eggshell, aqua, sky-blue, beige, etc. These are mixed, low intensity, or low saturation colors, respectively, as will be described later.

It is interesting to look at the light produced by different kinds of light sources through a diffraction grating or with a spectrophotometer. This allows us to observe what wavelengths are present and in what proportions. The light from a standard light bulb shows all the spectral colors, but the blue and violet are less intense than in sunlight or the light from a halogen lamp (Figure 6.3). If one decreases the current flowing through the light bulb, the filament becomes colder and the light emitted becomes dimmer and appears more yellow. When this light is analyzed with the spectrophotometer, it is evident that violet and blue have disappeared from the spectrum, explaining why the light appears to be more yellow.

Figure 6.3. Emission spectra from sunlight, a fluorescent ceiling light, an incandescent 25 W light bulb, and a 50W halogen bulb. Data from Gilbert and Haeberli, American Journal of Physics, submitted.

On the other hand, if you use a diffraction grating or a spectrophotometer to analyze the light coming from a neon discharge tube, you will find that only certain colors are present. You see many red lines, two or three orange lines, a yellow line, two faint green lines, and a few faint blue lines.

COLOR SENSITIVITY OF THE EYE

The human eye and its components were discussed at length in Chapter 4. The light-sensitive retina, found at the back of the eyeball, contains cone and rod cells. The *rods* provide low resolution, peripheral vision and function well even in dim light, while the *cones* provide color, high resolution, central vision and can only function in bright light. You can do an interesting experiment: make the lighting in a room very very dim, and wait until your eyes get used to the dark (dark adaptation). Now you can see things again but you will see no colors!

We will focus on the *color sensitive cones* because their functioning explains the perception of colors and color mixing. Studies on color vision indicate that there are three types of cones: let us call them Type I, II, III. They are sometimes called blue, green, and red sensitive cones, but this is not accurate: blue, green, and red are not the only colors to which these cones are sensitive. The sensitivity curves of the three types of cones are shown below.

The *sensitivity curve* captures how the neural response changes in intensity in response to changes in the wavelength of incoming light (when the intensity of the light is kept constant). A different neural response means that a different number of nerve pulses per second is transmitted by the cones in the retina to the neuronal axons in the optic nerve and the visual cortex at the back of the brain. In Figure 6.4, the three types of cones are labeled here with Roman Numerals and not with usual color names in order to stress an important point: *all three types of cones are sensitive to a broad band of wavelengths*, i.e. they are sensitive to many colors. Cones of Type I are sensitive to violet, blue, and green light, Type II are sensitive to blue, green, yellow, orange, and red light, and Type III to green, yellow, orange, and red light.

Figure 6.4.

The information that cones send to the optic nerve is simply a succession of electric pulses. The *rate* at which pulses are sent *depends both on the intensity of the light and its wavelength.*

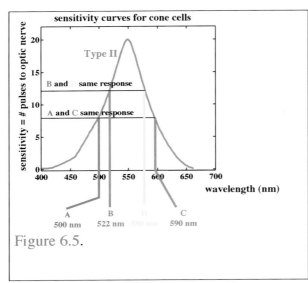

Figure 6.5.

Consider four different colors, as reported in Figure 6.5: A (cyan, with a wavelength of 500 nm), B (green, 522 nm), C (orange, 590 nm) and D (yellow, 580 nm). From the sensitivity curves of Figure 6.4, you can read the responses of Type II cones. Notice that the responses are identical for A and C and for B and D. *If you only had Type II cones, you would not be able to distinguish cyan from orange or green from yellow.* By having the **three types of cone cells**, the brain combines the responses from each type and is able to interpret *many more colors.*

Figure 6.6.

The pink dots in Figure 6.6 indicate the response of all three types of cone cells: both colors A and C generate a response of 8 on Type II cones, but at the same time, A generates only 3 on Type III cones while C generates 17. In addition, A also stimulates a small reaction on behalf of Type I cones. No such reaction is present for C. The brain receives a set of these three numbers: the responses of cones of Type I, II, and III. Processing this information, the brain can therefore distinguish A and C. This interpretation takes place instantaneously, as you can see and distinguish the colors as soon as you see them!

Most of the brain volume is used to interpret visual input. This tells you how complicated it must be! Despite the complexity of the system, there is a problem associated with having only three types of cones: some ambiguities occur.

Figure 6.7.

Figure 6.8.

In Figure 6.7 is a good example of the ambiguity generated by having *only* three types of cones. Consider two lights of wavelengths 485 nm and 652 nm. These are green and red lights. If you project these two lights onto a white screen (you *add* them), the brain will interpret them as the *sum* of the responses of all 3 types of cones. The response of Type I is negligible. For type II is 5.5 and for type III is 3 + 3 = 6 (see the pink dots in Figure 6.7).

Let us now look at a third light of wavelength 570 nm, which is yellow (Figure 6.8). This generates a response of 16.5 and 18 on Types II and III, respectively. These responses are identical to the ones obtained above from the sum of green and red, only 3 times more intense: $5.5 \times 3 = 16.5$ on Type II and $6 \times 3 = 18$ on Type III. That is why when you add red and green lights the eye perceives the resulting light as yellow even if no yellow wavelengths are present!

The *mixed yellow* appears dimmer than the spectral yellow, 3 times dimmer to be exact! This is because at the red and green wavelengths, both types of cones are less sensitive than at the yellow wavelength. Therefore the pink dots in Figures 6.7 and 6.8 are lower on the vertical scale for both the green and red lights than for the yellow light. Analyzed with a spectrograph, the mixed yellow has two wavelengths (485 and 652 nm) and the spectral yellow has only one wavelength (570 nm). Despite this spectroscopic difference, *the eye cannot tell the difference between one light of 570 nm and the sum of two lights of 485 and 652 nm*! Perhaps if we had 4 or more types of cones we would not have such ambiguities in our vision. On the other hand, more types of cones would require more volume of the brain to be used for vision, and this might have a negative effect on our other skills. Even worse, if we did not have these ambiguities, we would not be able to mix colors as easily! Then we could not have such a simple coding system for RGB color TV, computers, and projectors, and we would have much more difficulty mixing pigment colors.

Cats have very poor color vision because they have very few cone cells. Their eyes are optimized in several ways for night vision. Since cone cells only function in bright light and cats are nocturnal animals (more active at night), they have evolved to have a majority of rod cells that work well in dim illumination. Other differences between our eye and a cat's eye are summarized in Figure 6.9.

A cat's eye has *vertical slits* as pupils that can open up much more (at night) and more easily than the radial and circular iris muscles of a human's circular pupil. A larger aperture has higher luminosity because more light can go through the lens but also has the disadvantage of a small depth of field. The lowest f/ (fastest lens, highest luminosity) for camera lenses is f/1, for the human eye f/2.4, for an owl, another night predator, f/1.3, and for a cat, the *f number is f/0.09*. The cat also has a *tapetum* (Latin for carpet), a layer of reflecting cells at the back of the retina. The retina is semitransparent and it does not absorb all the light that illuminates it. Part of the light is transmitted through the retina and is simply lost. Night animals have the tapetum to reflect that portion of the light back into the retina, as indicated by the arrows in Figure 6.9. The tapetum is located immediately behind the retina, and therefore the reflected rays do not go out of focus much, allowing the rods in the retina to have a second chance to detect them.

cat's eye is optimized to see in dim light

- has small f/ (vertical slit instead of round pupil, for larger lend aperture D)
- has a **tapetum** to reflect light back to the retina
- has a **majority of rod cells** (few cone cells)

fast camera lens   f/1
human eye   f/2.4
owl eye   f/1.3
cat eye f/0.09

tapetum

Figure 6.9.

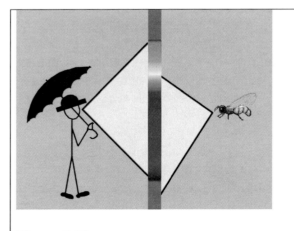

Figure 6.10.

Many snakes are thought to see infrared light but this is not true. Snakes do sense the heat irradiated (infrared) by their prey at night through their tongue, but they do not see it. One exception is the rattlesnake, which has an infrared pinhole system that allows it to actually see infrared light. Dogs and flies see colors just as humans. Bees do not see red but do see ultraviolet light, which humans cannot see. To attract bees, some flowers have structures on their petals that are only visible in the ultraviolet, and we cannot see (Figure 6.10).

Many other insects perceive UV light, as do birds, lizards, turtles, and many fish. All these animals have a fourth type of cone in their retina, which we humans completely lack. We do not have, by any means, the most sophisticated eye in the animal kingdom!

A more direct way of studying how our eyes perceive color is to analyze the addition of lights, or *additive color mixing*. This is described below and in more detail in Chapter 7.

Before starting to add lights, let's introduce a few criteria and definitions that will be useful to understand how colors are mixed.

PHYSICAL AND PSYCHOLOGICAL COLOR

The physical measurable color and the perceived, psychological color are different. An example of *physical color* is spectral yellow in sunlight or from a sodium lamp. This is a color to which only one wavelength is associated, or a *spectral color*.

The *psychological color,* or *hue*, can be *a single wavelength or a superposition of different wavelengths*. For example, think of yellow: the color we see can be a spectral yellow with a dominant single wavelength of 570 nm or the superposition of red and green lights with two different wavelengths, 485 nm and 652 nm. It could even be a broad-band spectrum that appears to have the same color as 570 nm spectral yellow. All these yellows are *metamers*: they have different physical origins but generate the same perceived color for the normal eye.

Before we can discuss color further, therefore, we need to establish an accurate definition of color.

DEFINITION OF COLOR: HUE, SATURATION AND BRIGHTNESS

The physical parameters of light are not sufficient to accurately describe the color we see: a psychological response to the physical stimulus generates the visual sensation. Color is therefore best described by *psycho-physical parameters*, which we will call hue, saturation and brightness. Here all three, and their synonyms and antonyms are defined. Given any color, a matching color that appears to the normal retina identical to the color in question, is obtained by combining light of a single wavelength with white light. The wavelength of the matching color is called *dominant wavelength*. The corresponding attribute of visual sensation is called the *hue* (or simply *color*). The fraction of white light is called *saturation*. Saturation is also known as *purity*. High saturation colors require a small fraction of white in the matching color, low saturation colors require a large fraction of white. These are pale pink, sky blue, pale yellow, beige, and all those colors commonly called "pastels". The adjectives saturated and pale are therefore antonyms. To match those colors, the matching color requires a large fraction of white. Some authors define saturation in paints and pigments to be the parameter related to the amount of black, but we do not see a reason for such distinction. *Brightness* (also known as *lightness*) is the parameter of a color according to which an area appears to emit more or less light. The adjectives bright and dark (or light and dark) are therefore antonyms. In the literature there is a preference to use the term brightness for luminant sources, and lightness for illuminated surfaces.

The hue, saturation and brightness are considered psychological parameters by other authors, and are deemed to be non-measurable. The computer industry, however, has made the decision of adopting these three terms (abbreviated HSB) to quantitatively and reproducibly describe colors. We therefore think it wise to adopt these three terms that, although not perfectly in line with previous physics publications, are in good agreement with every day language and with current technology.

Psychologically, pure yellow (0% white) appears to have *"low saturation"* and seems to be lighter, while red, green and blue appear to have *"high saturation"*, they seem to be darker and fuller colors compared to yellow. This is due to the higher sensitivity of the Type II and Type III cones in the yellow region (either single wavelength or broad-band yellow at 500-700 nm), compared to other wavelengths. There is no physical correspondence to this impression, which is entirely due to the physiological response to light and colors, not to the actual intensity or saturation of the physical color observed.

The *intensity* of the *physical color* corresponds to the *brightness* of the *psycho-physical color*. The perceived, psychological brightness cannot be measured. It is proportional to the logarithm of the intensity, but it varies dramatically from person to person—even for the same person depending on the surrounding illumination and on the *adaptation* of the eye to both color and light intensity. If the eyes are dark-adapted, that is if they've been in darkness for a few minutes or longer, when the light is turned on again, a dim color may appear brighter than it does when the same eyes of the same person have not been dark-adapted. The physical intensity of that color, which is a measurable quantity, remained unchanged in both conditions.

The perceived brightness of a color also depends on the surrounding illumination: a snowball indoors is perceived as white and a piece of coal outdoors under sunlight illumination is perceived as black, even though the snowball indoors scatters much less light than the coal outside! Brightness depends on the environment surrounding the object being observed. In bright sunshine a black object can still reflect a large amount of light, but since everything else around it reflects more, it is perceived as black. This effect does not only take place under intense sunshine. Observe the checkerboard in Figure 6.11.

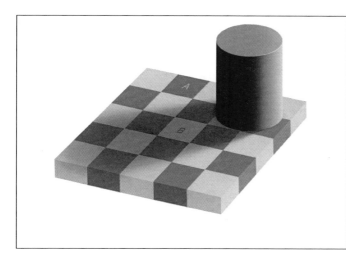

Figure 6.11. **The checker shadow illusion**
The squares marked A and B are the same shade of gray. To check this cover the figure with a piece of paper with hole punched to show only the areas A and B. Checker B appears much lighter than checker A because it is surrounded by darker shades of gray. You can find this and many other similar illusions on Edward H. Adelson's web site at MIT: http://www-bcs.mit.edu/gaz/

LIGHT INTERACTION WITH OBJECTS IN THE WORLD
The light illuminating objects can be:
- absorbed
- specularly reflected
- scattered or diffusely-reflected (these are synonyms)
- transmitted and refracted
- or combinations of the above

If light from a light bulb illuminating an object is completely *absorbed* by that object, no light is reflected and the object appears black. If only some wavelengths of light are absorbed and others are *scattered*, the object will appear *colored*. If light is *specularly reflected*, and the object is an aluminum or silver mirror, the light does not change color at all and is reflected in one very specific direction, defined by the law of reflection (i = r, as described in Chapter 2). In this case, the virtual image of the object can only be seen from a very specific position. Conversely, when light from the light bulb is *scattered* by the object, it bounces back in all directions and the object can be seen from any angle. If the object is transparent, light will be also *transmitted and refracted*. The most common situation is to have a *combination* of absorption and scattering of different wavelengths. This generates the colors of all objects surrounding us. If the objects are shiny, they are also specularly reflecting a portion of the light. Let us now look at the specific details of these situations.

SCATTERING OR DIFFUSE REFLECTION
Scattering (or diffuse reflection) takes place when light illuminates an object that does not have a perfectly smooth and flat surface. We have seen that in specular reflection, such as reflection from a mirror or a shiny smooth object, the angle of incidence is equal to the angle of reflection. In diffuse reflection this is not the case. In fact, an object can have any kind of surface roughness, texture, or finish and we can usually see it from all angles. Light illuminating any non-mirror object is *scattered in all directions* away from the object. The scattered light rays are then visible from all angles, all around the object.

In most cases, the conditions needed for specular reflection are satisfied at the microscopic level, but there are so many different microscopic surface orientations on a rough, non-mirror-like surface that the end result is light bouncing off in all directions.

For simplicity's sake, we can divide all the colors in white light into three spectral ranges: blue, green, and red, as shown in Figure 6.2. We are therefore neglecting the spectral orange, violet, and blue-green or cyan. This is an oversimplification in some cases, but it is useful and can be applied to most real-life situations and produce accurate results. In the examples below and throughout the remainder of this book, we will use this simplified spectrum.

A *black* object, such as a piece of lava-rock, *absorbs most of the light intensity* when illuminated by white light. A very small amount light is scattered, and from that small amount we can tell the shape and size of the rock. Our eye can see the small amount of light, perceive its very low brightness compared to the colors of the surrounding environment and to the illuminating light, and interpret the color as black. All colors in white light (red, green, and blue) are absorbed equally by the black rock, and a small proportion of each one, in equal amounts, is scattered.

A *white* piece of paper or a snowball behaves in a very similar way: all colors are equally reflected. The difference is that in this case, *no light is absorbed. All the illuminating light is scattered.* The blue, green, and red components are reflected in equal amounts, our eyes see the high brightness, and both the paper and the snowball appear white.

All colored objects absorb certain wavelengths and scatter others. The apple in Figure 6.12 appears red because, when illuminated by white light, it absorbs green and blue, and scatters red light only.

Figure 6.12.

A lemon appears yellow because it absorbs blue light and scatters red and green light. Our eyes perceive these red and green wavelengths, we add them together, and interpret the lemon's color as yellow. We can also think about the color perceived in another way: *an object appears as the color complementary to the color it absorbs.*

Figure 6.13.

In other words, a lemon appears yellow because it absorbs blue, the complementary color to yellow. Furthermore, the lemon appears shiny on the surface (see Figure 6.13) because a small portion of the light (white light!) is specularly reflected. The same goes for the apple of Figure 6.12.

The *complementary colors* are:

yellow ⇔ blue
magenta ⇔ green
cyan ⇔ red

They appear on opposite sides of the diagram in Figure 6.14.

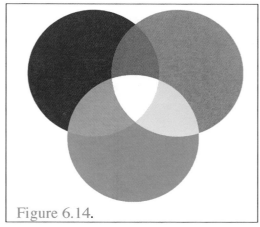

Figure 6.14.

It is interesting to note that Descartes in 1630 attributed the color of an object to a change in the illuminating light, when this is reflected by the object. This is exactly what we now know is the correct interpretation of color: a red apple is red because it absorbs all colors in the illuminating light except for red, and scatters red. The illuminating and reflected lights, therefore, are different, exactly as Descartes intuition predicted.

Scattering also depends on the size of the scattering illuminated particles. If scattering particles are larger than the wavelength of light, the scattering is the same for all wavelengths (all light colors). If the particles are smaller, the scattering is wavelength-dependent. Let us see a few cases in which scattering produces interesting effects.

Water is almost completely transparent. But then, why does *water* in a fountain or a *waterfall* appear *white*? This phenomenon is an effect of scattering. The water particles, that is, the droplets in the waterfall, are much larger than the wavelength of visible light (400-700 nm). Therefore, light scattering is *wavelength independent*. This is also called *Mie scattering*. If white light is illuminating the waterfall, white light is scattered in all directions by the water droplets. Since all colors (wavelengths) composing white light are scattered equally, we see the waterfall as white. The same wavelength independent scattering is responsible for the *whiteness of the clouds*. The ice crystals and water droplets that form clouds are larger than the wavelengths of light and so they too appear white.

More about scattering: *why is the sky blue*? Air, with very good approximation, is transparent, so why does a lot of air (the sky) appear blue? In this case, the air particles (76% nitrogen molecules and 21% oxygen molecules) are much smaller than the wavelength of light and this causes the scattering to be *wavelength dependent*. This is also called *Rayleigh scattering*. The amount of light undergoing Rayleigh scattering is proportional to $\frac{1}{\lambda^4}$, where λ is the wavelength of light. Therefore, blue light, which has the shortest wavelength (400-500 nm), is scattered more than green (500-600 nm) and red light (600-700 nm), and the sky appears blue. Although blue is scattered more, all wavelengths are scattered: this is why the sky is light blue (in fact sky blue = unsaturated blue = white + blue).

Sometimes the sky is very light blue, other times it is a much more saturated blue. Wavelength-independent Mie scattering from water vapor desaturates the sky color. In dry climates the sky is always more saturated. Figure 6.15 clarifies the geometry of direct sunlight and scattered light from the sky.

Figure 6.15. Formation of the blue sky and red sunrise or sunset by scattering in the atmosphere (adapted from K. Nassau, The physics and chemistry of color, Wiley, New York, 2001)

Imagine light from the sun coming towards you. Along the way it encounters air molecules in the atmosphere and bounces off of them at all angles. Then, the blue light, which was scattered more than green and red at the first bounce, continues to bounce off other air molecules again and again. The whole sky becomes illuminated homogeneously as blue. In the absence of an atmosphere to scatter light, the sky would not be illuminated at all and it would appear black. This is the case for the Moon, which does not have an atmosphere. On the Moon, with no light scattering from air molecules, areas that are directly illuminated by the sun (in line of sight of the sun) are extremely bright and the shadows are completely dark.

Fog and mist, as clouds, are formed by water droplets larger than the wavelength of light (usually 1000 nm or larger), therefore they scatter wavelength-independently, and appear white. Manufacturers of automobile fog lamps, however, sometimes make these produce yellow light, so that less scattering occurs from the smaller droplets that are also present.

Why is the ocean blue? It is by no means true that the ocean is blue because it reflects light from the sky. This can easily be proved on the surface of still water (e.g. in a swimming pool) with an external source of light and a polarizer filter. It is also not true that blue in water and ice is a result of scattering. Although some scattering is certainly present, the amount of scattering-generated blue is orders of magnitude lower than the color generated by water absorption. The answer is simple: the ocean is blue because water is blue, or cyan, to be perfectly accurate. The red part of the spectrum is absorbed by water, leaving blue and green (which are not absorbed) to be scattered. Not by chance, "aqua" is a blue-green color (blue-green = cyan and is complementary to red). Water absorbs red, therefore appears cyan. The same is true for the pale cyan color of bulk ice, e.g. in glaciers or icebergs. A green color in water or ice usually indicates the presence of algae, which contain a green pigment (chlorophyll).

When the background is white, e.g. on white sand, or in a white marble fountain, or simply in a white bathtub, water appears cyan, because it absorbs red. Figures 6.16 and

6.17 show these examples.

Figure 6.16. Colors of the Pacific Ocean from the Huahine island in French Polynesia. At the center of this photograph there is a white sand bank underwater, the shallow water above it, therefore, appears an unsaturated cyan, that is white (from the sand) and cyan (from the water itself). Outside the reef, where the water is deeper and rougher, the wind produces white caps, foam and isolated air bubbles, which scatter light and keep it from penetrating into the water. As a result the water color is darker blue. In the foreground of this photograph the water is also deeper, but not rough and bubbly, it therefore has an intermediate brightness.

Figure 6.17. Water in the Barcaccia fountain in Piazza di Spagna, Rome, Italy, designed by Pietro Bernini, father of Gian Lorenzo. Shallow water against the white marble background appears cyan because water absorbs red.

In summary, from all these examples we conclude that *the color of an object depends on many different aspects*, including:
- the wavelength and intensity of the illuminating light;
- the wavelength of light scattered or transmitted by the object;
- the color of surrounding objects;
- and also on the absorption or scattering of substances in the path of light.

Chapter 7
ADDITIVE COLOR MIXING

Additive color mixing works for *mixing lights*. The rules of additive color mixing apply to color TV, computer monitors, projectors, and stage and disco lights. These rules DO NOT apply when mixing pigments and

paints. The three projectors shown here all emit white light. Each projector is illuminating one filter. The color each filer transmits onto the screen is the same color as the filter is. The red filter transmits red light, the green transmits green light, etc. We can now mix these three colored lights by projecting them onto the same location on the white screen. By overlapping them, we can obtain almost any colors. On a TV or computer monitor, instead of overlapping the tree colors, they are positioned side by side in small areas called "pixels". The pixels are small enough that from a normal viewing distance the eye perceives the additive mixture of red, green and blue. Look at you TV or computer screen with a magnifier and you will see the individual pixels. Let us start with the additive primary colors.

PRIMARY COLORS

The three filters in front of the three projectors of *identical intensity* are red (R), green (G) and blue (B). If all three colored lights are projected on a white screen, partly overlapping, they produce the additively mixed colors displayed in Figure 7.1. As shown, adding blue light to red light produces magenta (M) (B + R = M), while adding B + G = cyan (C) and adding R + G = yellow (Y). Where R, G and B overlap, white is obtained. R, G and B are therefore three *additive primary colors*. The definition of primary colors is:

- *one primary cannot be matched by a mixture of the other two.*
- *all three, when combined in equal amounts, produce white.*

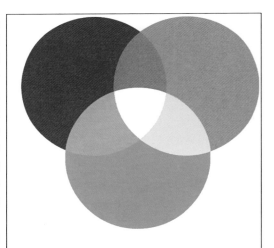

Figure 7.1. The most frequently adopted additive primary colors are R, G and B.

There are many other combinations of three colors that correspond to this definition. An example can be seen in the above diagram: C, M and Y. These could also be primaries since each color cannot be made by a mixture of the other two and, when combined in equal amounts, they produce white. As we will see in more detail later (Figure 7.9), however, when mixed, they can only generate a limited number of colors. Therefore, they are not the most useful additive primaries.

Two *complementary colors* are *two colors that, when additively mixed, produce white.* Additively mixed implies that we are adding lights of these two colors, as with the above projectors and colored filters (not paint!). Note in Figure 7.1 that the complementary colors appear on opposite sides of the diagram, and that the complement of each primary color is the combination of the other two primaries.

ADDING PRIMARY COLORS

It does not seem possible, at a first sight, that we can mix R, G, and B to obtain the two colors reported in Figure 7.2: chocolate brown and cyclamen pink. We will describe in detail how to do this quantitatively. For now, it will suffice to say that there are accurate combinations of R, G, and B that produce these two colors. These are: 66% R, 33% G, and no B. Then to obtain brown, we must also decrease the intensity of the light. Pink is a mixture of approximately 40% R, 30% G, and 30% B.

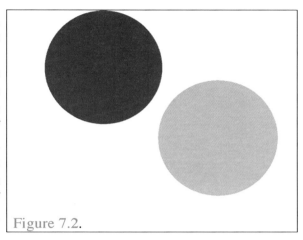

Figure 7.2.

In fact, *almost any color can be obtained with a combination of R, G and B* lights.

Let us illuminate a white screen using a projector with a light of an arbitrary color (stimulus S). In Figure 7.2, S is brown or cyclamen pink. To do this, we can place on the screen a colored sheet of paper and illuminate it with white light, or have a colored filter in front of the white light projector. Let us now illuminate an empty part of the white screen in Figure 7.2, with R, G and B lights, all pointing to the exact same location, so they overlap. If we vary the intensities of these three lights, we can obtain a match for brown or pink. A match of the color can be represented by the equation:

S = aR + bG + cB

In this equation, a, b, c are the intensities of R, G, and B lights, respectively, that are needed to match the stimulus color. The units or range of values used for the intensities (a, b, c) are of no interest. The fractions of R, G, and B that make up the light are the only numbers that matter when defining the color of the light. We can calculate what fraction r of the light is red:

$$r = \frac{a}{a + b + c}$$

And similarly what fraction g is green:

$$g = \frac{b}{a + b + c}$$

These two numbers, r and g, define the mixture of colors projected on the left spot.

In the cyclamen pink example we have:

S = 250 R + 170 G + 200 B

The numbers reported here for a, b, and c (250, 170 and 200) are the intensities of R,G and B used on the computer to generate the cyclamen pink in Figure 7.2, each of which can vary between 0 and 255. The total Intensity I is the sum of the intensities for R, G and B:

I = a + b + c = 250 + 170 + 200 = 620 units.

Remember, the units in the intensity are irrelevant because the only numbers that matter to define the color are the ratios, or the fractions, of R, G, and B. These ratios are unitless. On a computer with 8 bit color images, the values of a, b, and c vary between 0 and 255. There are therefore 256 (= 2^8) intensity levels for R, for G, and for B. You can try the numbers above on your computer and verify that you really do obtain cyclamen pink. The ratios we need to calculate are:

The fraction of red $r = \dfrac{250}{620} \approx 0.40$

The fraction of green $g = \dfrac{170}{620} \approx 0.28$

The fraction of blue $b = \dfrac{200}{620} \approx 0.32$

Notice that r + g + b = 0.4 + 0.28 + 0.32 = 1

This is always true! *The fractions r, g, and b always add up to 1*. Therefore, we do not need to calculate all three fractions. Calculating the fraction of red and the fraction of green will be sufficient because the fraction of blue will always be the remainder (b = 1 – r – g).

We can now plot the fraction of red on the horizontal axis and the fraction of green on the vertical axis of a diagram such as the one in Figure 7.3. This diagram is called the chromaticity diagram or the *color triangle*.

THE COLOR TRIANGLE

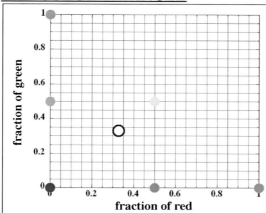

Figure 7.3. **The color triangle**
The fraction of red and green may vary between 0 and 1. The fraction of blue does not need to be plotted because it is always the complement to 1 of the other two. If the fraction of green is 0, the colors on the horizontal axis are: B when r = 0, R when r = 1 and M when r = 0.5 (mid-point between R and B). On the vertical axis (r = 0), we will have G when g = 1 and C when g = 0.5 (mid-point between G and B). Y is between R and G, or when r and g = 0.5. White is at the point where both r and g are 0.33. The color triangle on the left shows all these points (R, G, B, C, M, Y, W) and the corresponding colors.

We can find the coordinates of any mixed color on the color triangle. The color triangle rule is the following: when *adding two colors, the resulting color is always on the line joining the two colors*.

This is why C is in the mid-point between B and G, M between B and R, and Y between R and G. The color triangle in Figure 7.4 provides a simple way of mixing additive primary colors to obtain a desired color, or to anticipate exactly what color will result when two colors are mixed in known proportions.

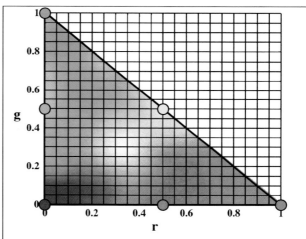

Figure 7.4. The color triangle here is filled with all the colors which can be obtained by mixing R, G and B.

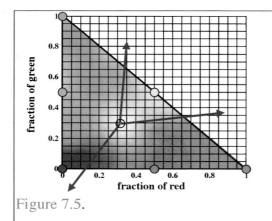

Figure 7.5.

Any mixed color is the sum of the corresponding spectral color plus white. A high saturation color contains no white, while a low saturation (pastel) color contains lots of white. In Figure 7.5 along any line starting from white and going towards the edges of the triangle, the hue of a color does not change. The only thing that changes is the saturation. Saturated colors are near the edges of the triangle while low purity colors are near the center.

Once the fractions of red and green of a color are identified, all we need to do to find the exact hue of the color is graph the point (with coordinates r and g) on the color triangle, draw a line from white to that point, extend the line until it reaches the edge of the triangle, and identify the color where the line crosses the edge. We then identify the saturation of the color by measuring how far the point is from the edge.

Take the example of cyclamen pink. We found that r = 0.4 and g = 0.28. We graph the point with these coordinates on the triangle in Figure 7.6, and report it as a white circle, extend a line from white to this point, and then continue the line until we reach the edge. The color at the edge (the corner in this case) is red. The point corresponding to our color is much closer to white than to red, we therefore conclude that the color is a low saturation red. This is known outside of physics as pink. If the distance from white to pink is 5-times smaller than from white to red, the pink is in fact $1/5^{th}$ red and $4/5^{th}$ white.

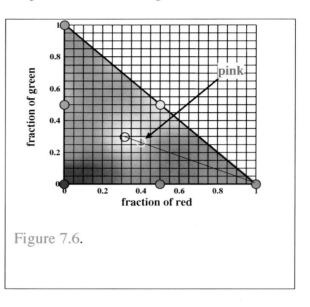

Figure 7.6.

When mixing two colors, we must first identify them on the color triangle. Then we must draw a line between them. If the two colors were mixed in equal amounts, the resulting color is at the mid-point between the two original colors. If there was more of one color than the other, the resulting color is closer to the more abundant of the two.

According to the color triangle rule, when mixing two colors, the resulting color lies along the line connecting the two. In Figure 7.7 we mix 1 part of high saturation orange and 2 parts of high saturation green, the resulting color lies 1/3 of the way from green and 2/3 of the way from orange along the line connecting the two colors, and the resulting color is a high saturation lime green (thicker white circle). Similarly, if we mix 2 parts of Y and 1 part of C, we obtain low saturation lime green. Imagine this point on the color triangle.

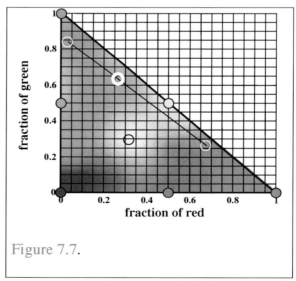

Figure 7.7.

By now, it should be clear why R, G, and B are usually chosen as the additive primary colors: RGB provide the largest mixing potential. In other words, the largest number of colors can be obtained from mixing RGB primaries: all the colors in the color triangle. *If we chose red, blue and yellow* as additive primary colors, we could only obtain the colors within the triangle enclosed by the three points where these three colors lie on the color triangle, as shown in Figure 7.8. We could obtain R, O (for orange), Y, W, B, M and all the colors intermediate between these, but *we could never obtain G or C.* This is shown below in the diagram on the left. *A choice of C, M and Y would be even more constricting as we could never obtain red, or blue, or green.* This case is shown in Figure 7.9.

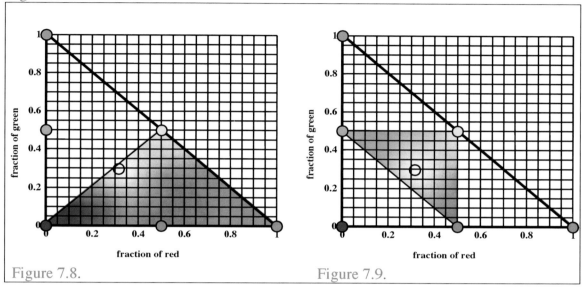

Figure 7.8. Figure 7.9.

LOW INTENSITY COLORS

Are *all* colors represented on the color triangle? Where are the olive greens, the grays, the browns, beige, and black? These colors are indeed present on the triangle, but since *the color triangle does not represent the color intensity*, these low intensity colors are not immediately evident. Olive green is nothing but low intensity green, while brown is low intensity orange, gray is low intensity white, and black is the lowest intensity white. Black can actually be the lowest (or zero) intensity of any color. Black could therefore be anywhere on the color triangle.

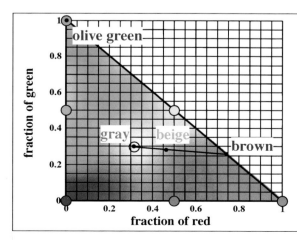

Figure 7.10. Olive green, brown, and gray are shown as dark dots in the color triangle. They are in the same positions as green, orange, and white since these are only low intensity colors and the color triangle does not reveal intensities. Beige is low saturation brown and it lies along the line between white and brown (or orange).

For the above colors the RGB intensities on a 0-255 level basis are the following:

Green = 0R + 255G + 0B
Olive green = 0R + 100G + 0B
White = 255R + 255G + 255B
Black = 0R + 0G + 0B
Gray = 123R + 123G + 123B
Orange = 255R + 123G + 0B
Brown = 123R + 62G + 0B
Beige = 223R + 162G + 100B

The colors chosen for typing the color names on the diagram above were obtained using these accurate values. Please try them on your own computer: any graphics program (Adobe Photoshop®, Adobe Illustrator®, PowerPoint®, Canvas®, etc.) will allow you to enter the RGB numeric values you desire when using a "color picker" to select a color.

SPECTRAL COLORS

Not all the spectral colors can be matched by mixing RGB. *The colors obtained mixing RGB primaries are less saturated than their corresponding spectral colors* (dominant wavelengths). This is particularly evident in the *blue-green* region, also called *cyan* or *turquoise*.

Let us project on a screen in spot #1 a combination of RGB and on spot #2 spectral cyan (470 nm). Let us try to match the color of spot #2 with a combination of RGB in spot #1. As shown in Figure 7.11, in spot #1, we would project blue and green, with wavelengths 420 nm and 550 nm, respectively. We need to use our imagination here, because the color in spot #2 was produced on a computer that works additively and is therefore not as saturated as the true spectral cyan with 470 nm. Please imagine that the cyan in spot #2 has a much higher saturation than the mixed cyan of spot #1.

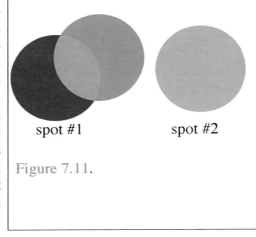

spot #1   spot #2

Figure 7.11.

To obtain a perfect match, we need to produce the same response in the cone cells of our eye. We will find that the only way to obtain a match is to shine some red light on spot #2 and then match the new color with blue and green light on spot #1. The reason for this is clearer if we look again at the sensitivity curves for the cones, and their responses to the stimuli on spots #1 and #2. These are reported in Figures 7.12 and 7.13.

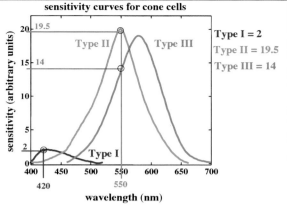

Figure 7.12. The responses of the three types of cones to the stimulus of spot #2: spectral cyan with 470 nm wavelength.

Figure 7.13. The responses of the three types of cones to the stimulus of spot #1: a mixture of primary G and B with 550 and 420 nm wavelengths, respectively.

If we try to obtain the same result as from C by adding B and G, we must first adjust the B intensity to give the correct type I response (about 1 unit in Figure 7.12). We then add enough 550 nm light to get the desired type II response of 4 units. Since 550 nm *also* stimulates type III cones a lot a lot (about 3/4 as much as type II, in fact at 550 nm for 19.5 units on type II we also get 14 units on type III), however, this means we have added too much red, namely about 3 units (3/4 of 4) (Figure 7.13).

The Adding R to B+G is equivalent to have added white, so the resulting color is less saturated than the spectral cyan. The only way to match spectral cyan with a mixture of B and G, therefore, is to add red to cyan, and then match it.

Adding red to the spectral cyan means to *desaturate* the cyan. Since red is the complementary color to cyan, the addition of red is equivalent to adding some white, and produces a less saturated color. Adding red to the spectral cyan we were trying to match is equivalent to adding *negative red* to the mixed color. This is physically impossible: since there was no red to begin with in the B+G mixture, it is not possible to subtract red from it. However, a negative fraction of red can indeed be plotted on the color triangle, or, rather, *outside the color triangle*, as shown in Figure 7.14.

Figure 7.14.

The spectral cyan lies outside the color triangle. The color triangle defines the colors that can be matched by a mixture of RGB, but more colors do exist—they just can't be matched by RGB. Cyan is an example of these colors. Many other spectral and non-spectral colors lie outside the triangle. These colors have to be *desaturated* before they can be matched by a mixture of RGB primaries. You will never see an RGB mixed turquoise on TV or on your computer monitor that is as saturated as the spectral turquoise.

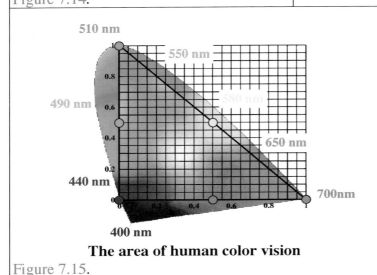

The area of human color vision

Figure 7.15.

The hues obtained mixing primaries are less saturated than their spectral corresponding colors, which are each associated with a single (dominant) wave-length.

As shown in Figure 7.15, most of *the spectral colors lie outside the color triangle*. Spectral red, green, and blue are at the edges of the triangle, while yellow, cyan and violet are outside of it. So are the purples.

The horse-shoe shape in the figure above represents *the area of human color vision*. This area is enclosed by the curved line along which all spectral colors lie (see corresponding wavelengths in the diagram) and the line between spectral violet and red. In summary, *the color triangle includes only the colors that can be obtained mixing RGB, but the eye can see more colors, including spectral colors and purples*.

NON-SPECTRAL COLORS

Purple and magenta, as well as all the magenta-purple hues, are *non-spectral colors*. There is no wavelength associated with them, because they are not part of the spectrum of white light. They are exclusively mixed colors. Magenta is the mixture of spectral red (700 nm) and spectral blue (440 nm), while purple is the mixture of spectral red (700 nm) with spectral violet (400 nm). The mixture of R+B in different proportions generates all the magentas, the reddish-magentas and the bluish-magentas. These hues can be obtained mixing RGB primaries (R+B), and therefore you can see them on TV or on a computer monitor. The purple hues cannot be mixed with RGB primaries (since they lie outside the color triangle).

Now that we described the color triangle and the spectral colors, let us look at the complementary colors again. Two additive complementary colors must produce white when added together. White must be on the line between the two colors. For example, B and Y are complements because when additively mixed they produce white. To be perfectly accurate, *B and spectral Y* are *complementary colors*, not B and Y, since the points B and Y are not equally distant from white, while SY and B are equidistant from W. We would actually need to add a lower saturation blue (sky blue, labeled SB in Figure 7.16) to yellow if we wanted to obtain white. As an aside, note that B and Y pigments do not produce white when mixed, as will be explained in detail in Chapter 8.

The line connecting blue and yellow passes through white. Other pairs of complements are shown in Figure 7.17, such as bluish-cyan and orange (BBG (or BC) and O), lime green and bluish magenta (YG and BM), green and magenta (G and M), but many more exist. Again, to be accurate, it is unsaturated green (UG) that complements magenta perfectly.

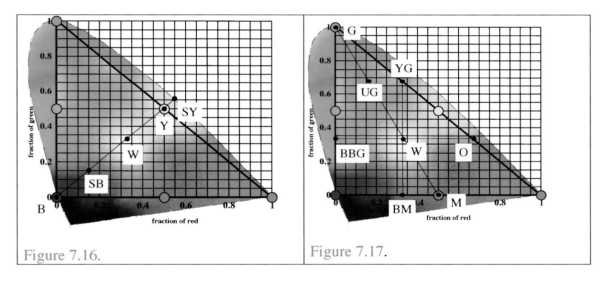

Figure 7.16.

Figure 7.17.

SUMMARY
Every color can be located on the color triangle by a three-step process. First match the color with three primaries R, G, and B. Second, find r and g:

r = fraction of red in the total intensity of light in the match;

g = fraction of green in the total intensity of light in the match.

Third, plot r on the horizontal axis and g on the vertical axis.

White falls at the point r = 0.33, g = 0.33, near the center of the triangle. The more saturated a color (i.e. the greater its purity) the further it falls from white. Spectral colors are the most saturated and fall on a curved line outside the triangle.

When two colors are added, the resulting color always falls along the line connecting the two. This idea is useful in the explanation of complementary colors—colors that produce white when added. Complements always fall on opposite ends of a line through the white point. The hue of a color can be specified by finding the dominant wavelength, or in the case of purples, the complementary wavelength of the color. Every color can be matched by a spectral color, or a purple, plus white. The smaller the amount of white, the higher the saturation.

ADDITIVE COLOR MIXING IN PAINTING

When painting and mixing oil paints, the rules of additive color mixing don't usually apply. We will describe in Chapter 8 how to mix pigments and paints with subtractive color mixing. Nevertheless, there have been a few extraordinary cases in art in which painters have attempted to paint additively.

Post-impressionists used a revolutionary technique to enhance color brightness and contrast. This technique was based on additive color mixing. Van Gogh wanted to achieve the effects of the art glass in the windows of gothic cathedrals. These windows have colored pieces of glass adjacent to each other, and they transmit colored lights exactly as our projectors through colored filters.

Figure 7.18. Self portrait with hay hat
Vincent Van Gogh (Dutch: 1853-1890)
Post-Impressionist

Van Gogh did not know about additive color mixing and mixing lights, but he was striving to achieve the brightness and color contrasts that only mixing lights can achieve. He invented the "*mélange solide*", French for "solid mixture," in which instead of mixing oil colors subtractively, he painted adjacent brush strokes with contrasting colors (Figure 7.18). He described it saying "je cherche à exprimer avec le rouge et le vert les terribles passions humaines"— the terrible human passions are expressed by red and green, not mixed but adjacent. Since the brush strokes in his mélange solide are large, much larger than the eye's resolution, the eye does not actually mix lights from adjacent colors, it perceives the striking contrast of very different and very bright colors, as in a gothic cathedral window!

If the regions of two different colors are much smaller, lights from the two different colors side by side are mixed additively by our eyes and create a third color. This resulting color is described as optically mixed—that is, it is mixed in the eye of the observer. When two colors are mixed with pigments by a painter (subtractively) they lose some of their intensity, while when the same two colors are mixed optically, they retain their intensity and they may appear brighter. Two Neo-Impressionists, *the pointillists* Seurat and Signac, explored this possibility. They experimented with small color regions: points or dots, as shown in their painting in Figure 7.19 and 7.20.

Figure 7.19. Chenal de Gravelines
Georges Seurat (French: 1859-1891)
Neo-Impressionist

Seurat studied the effects of different colors interacting with each other, and he tried to control their mixing by reducing the number of colors on his palette and applying pigments in very small spots. The Impressionists have often been interpreted as using "broken colors", but a patch of yellow in an impressionist painting is almost always yellow paint, not strokes of red and green laid side by side to achieve optical mixing. Seurat, however, did construct magenta from small spots of blue and red, and with his system, he could also vary the intensity of red, for example, by placing it near an orange dot, and determine how much the red will induce a perception of its complementary (cyan) in the adjacent color.

Figure 7.20. St. Tropez, le port
Paul Signac (French: 1863-1935)
Neo-Impressionist

Seurat called his techinque *divisionism* because he had divided the colors into their physical elements. His theory and practice, together with those of his followers, were soon described as Neo-impressionism and Pointillism. The subject was still Impressionist – ordinary people seen in every day activities, atmospheric conditions, immediacy of the scene – but presentation and technique were new.

The Neo-impressionists had color theory available to them — including the works of Herman von Helmholtz (whose *Manual of Psychological Optics* had by then appeared in French) and Wilhelm von Bezold (whose texts were published in French in 1876). Seurat and his colleagues had begun to feel that if they did not become more involved in the science behind their technique, they would not be able to explain the optical effects forming the basis of visual art and art itself would remain intellectually unsatisfying. Seurat's task was made easier by the work of the American physicist Nicholas Odgen Rood (N. O. Rood, *Modern Chromatics with Application to Art and Industry*, Chicago 1879), who was also an accomplished painter. Seurat was inspired by Rood's writings and color theory, and he imitated the behavior of light itself, performing additive color mixing when actually optically mixing pigments. However, he was incorrect in thinking that "adjacent dots of pure spectral colors could be mixed [in the eye of the observer] to obtain other spectral colors observed in nature, with a brilliance and luminosity [read saturation!] unattainable by conventional pigment mixtures." Scientifically speaking, Seurat's idea was flawed. Nevertheless, he has given us some of the most haunting images of European art. Another problem was with the contrast enhancement of adjacent complementary colors. We have now learned that lights of complementary colors produce white when mixed—*not* saturated colors.

The optical mixture of complementary colors kills color, as Signac himself later admitted: "Pointillage makes the surface more lively, but it does not guarantee luminosity, intensity of color, or harmony. The complementary colors, which are allies and enhance each other when juxtaposed, are enemies and destroy each other if mixed, even optically. A yellow and a blue, if juxtaposed, enliven each other, but yellow dots and blue dots make an aggregate which is grey and colorless." You can verify this effect very simply. Next time you go to the Art Institute of Chicago, if you stand in front of Seurat's "Dimanche après midi sur l'Isle de la Grande Jatte" (Figure 7.21) and are close enough to see the individual dots, all colors appear very bright and luminous (saturated). If you stand far away they all blend together, and the overall effect is a hazy gray patina covering the whole painting.

Figure 7.21. Dimanche après midi sur l'Isle de la Grande Jatte
Georges Seurat (French: 1859-1891)
Neo-Impressionist

Chapter 8
SUBTRACTIVE COLOR MIXING

Subtractive color mixing applies to overlapping colored filters, and color photography and is essential for understanding the mixing of pigments in paints or inks, and for color printing

FILTERS

If light of a given spectral distribution passes through a *filter*, the spectrum of the light is modified. The filter absorbs part of the light and transmits the remainder.

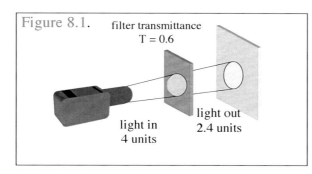

Figure 8.1.
filter transmittance
T = 0.6
light in
4 units
light out
2.4 units

The *transmittance* **T** of a filter is the fraction of the incident light transmitted through the filter. The filter in Figure 8.1 has T = 0.6. The filter absorbs 40%, and transmits 60% of the light. If the incoming light has an intensity of 4 units, then the transmitted light has an intensity of $4 \times 0.6 = 2.4$ units.

The transmittance of a filter is usually different for different wavelengths (colors) of light. The transmittance is only the same for all wavelengths if the filter is a *gray filter*, also called *neutral density filter*. Let us look at a colored filter.

Figure 8.2. Idealized spectra of colored light incident on a colored filter. The spectrum of the illuminating light is on the left, the transmittance curve of the filter at the center, and the spectrum of the transmitted light on the right. You should be able to decide what color the incident and transmitted lights are in these plots.

Once again we are using a simplified version of the spectrum, in which the range of wavelengths between 400 and 500 nm is referred to as blue, 500-600 nm is green and

600-700 nm is red, respectively (B, G and R). The real spectra are curved lines but they can be simplified as step functions, with vertical and horizontal lines, as in Figure 8.2. While this approximation is not *always* perfect to describe the behavior of light and color, it works in most cases and it is much simpler to use than the real spectral lineshapes. Examples of real filter curves are reported in Figure 8.3.

Figure 8.3. Transmittance curves of real filters. These are: R, G, B interference filters (top) and of C, M, Y interference filters (bottom). The filters used are commercial dichroic filters. Data from Gilbert and Haeberli, American Journal of Physics, submitted.

If you were not able to correctly identify the colors in the spectra of Figure 8.2, here is a complete explanation. The first spectrum was the incident light. This light has no intensity (Intensity = 0) in the B region of the spectrum and intensity I = 1 in both the G and R regions. The incident light is therefore a mixture of G+R lights. As described in Chapter 7, G and R lights combine to form Y light. Therefore, you know that the incident light is yellow. The filter transmits all light in the B and G regions (where Transmittance = 1) but no light in the R region (where Transmittance = 0). Since there was no B in the incident light, the only light that the filter can transmit is G. Therefore, the spectrum of the transmitted light has Intensity = 0 in both the B and R regions and Intensity = 1 in the G region, transmitting all the green light and causing the transmitted color to be green. The filter *subtracted* red from the incident light. It *absorbed* the R component from the incident light. We are now beginning to see why this is called *subtractive color mixing*.

SUBTRACTIVE PRIMARY COLORS

The most commonly used *subtractive primary colors are cyan, magenta and yellow*. Each one of these colors absorbs one additive primary, its complementary color.

Cyan absorbs red, magenta absorbs green, and yellow absorbs blue. When combined, *C, M, and Y produce black*. Again, these are not the only colors that can be chosen as subtractive primaries. Any other three colors that combined produce black are potential primaries, although the mixing potential—the number of colors that can be obtained mixing three primaries—is maximum with CMY.

This is why the computer industry has chosen CMY inks to be used in color printers, and the same choice was made by all manufacturers for color photography.

Figure 8.4 shows the absorption of the three subtractive primaries when CMY filters are used. Figure 8.5 shows how to obtain the additive primaries using combinations of CMY filters. In general, paints, inks, and pigments behave the same way with scattered light as filters do with transmitted light. Figure 8.6 shows the absorption of R, G and B filters, which also absorb their complementary colors (see Figure 6.14 for the definition of complementaries).

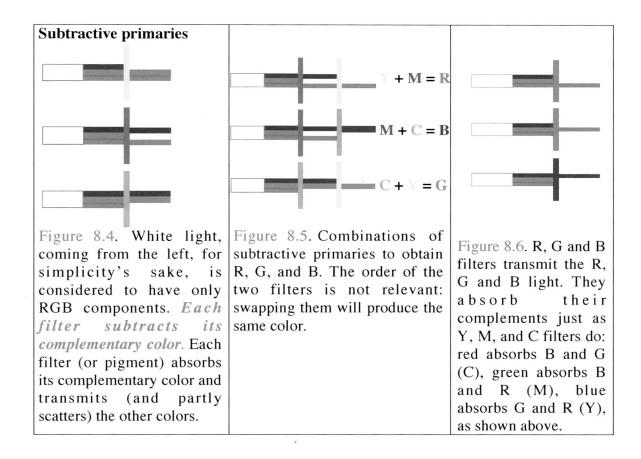

Subtractive primaries

$Y + M = R$

$M + C = B$

$C + Y = G$

Figure 8.4. White light, coming from the left, for simplicity's sake, is considered to have only RGB components. *Each filter subtracts its complementary color.* Each filter (or pigment) absorbs its complementary color and transmits (and partly scatters) the other colors.

Figure 8.5. Combinations of subtractive primaries to obtain R, G, and B. The order of the two filters is not relevant: swapping them will produce the same color.

Figure 8.6. R, G and B filters transmit the R, G and B light. They absorb their complements just as Y, M, and C filters do: red absorbs B and G (C), green absorbs B and R (M), blue absorbs G and R (Y), as shown above.

In subtractive color mixing, one often obtains colors that seem unexpected or surprising. The *hue and the saturation* of the light transmitted through two consecutive filters is not intermediary between the hues of the two filters. The same applies for pigments. You can look on the color triangle to see where the colors lie and compare that with the colors obtained anywhere in this chapter when overlapping filters or mixing pigments.

Combinations of overlapping filters produce the results shown in Figure 8.7.

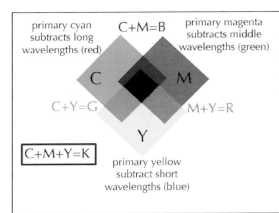

Figure 8.7. Three overlapping C, M and Y filters. Each two of these combined transmit an additive primary color (R, G or B), while all three do not transmit any light (K=black).

Notice from Figure 8.7 that the overlapping of red (or M+Y) and green (or C+Y) filters is black. Therefore,

Subtractively, R + G = Black
Additively, R + G = Yellow.

You see that subtractive color mixing is very different from additive color mixing! Mixing filters (or pigments and paints) is very different from mixing lights.

As discussed, overlapping filters and mixing pigments, paints, or inks is called *subtractive color mixing*. The term is slightly misleading because a filter does not subtract (absorb) a fixed amount of light but, rather, it absorbs a fixed *fraction* of the light incident upon it. The mathematical operation is not a subtraction. It is the **multiplication** of the incident intensity by the transmittance of the filter or the multiplication of the transmittances of two filters.

Let us look more in detail at a yellow and a cyan filter, overlapping to transmit green light, as shown in Figure 8.8.

Figure 8.8. White light, coming from the left, can be for simplicity's sake considered to have only RGB components. *Each filter subtracts its complementary color.* The cyan filter absorbs R (its complementary color) while transmitting B and G. Remember that

additively B + G = C. Therefore, the B and G lights transmitted by the filter form C. The yellow filter absorbs B and transmits G and R. Again, additively G+R=Y, therefore: a C filter transmits C light, a Y filter Y light. However, the combination of the two filters will only transmit G light. If we start from white light, we have R, G and B. The cyan filter subtracts R and the yellow filter subtracts B, leaving only G to be transmitted. (R + G + B) – R – B = G.

In Figure 8.9 are the transmittance curves and table for this example.

Simplified spectral ranges	T_{yellow}	T_{cyan}	$T_{Y\&C} = T_Y \times T_C$
B (400-500 nm)	0	1	0
G (500-600 nm)	1	1	1
R (600-700 nm)	1	0	0

Figure 8.9.

The resulting transmittance of the two filters can be graphically obtained by combining the transmittance curves of the two filters. In Figure 8.10 you can see that the region in which the two curves overlap (the dashed region) is the resulting transmittance. In this region neither of the filters had a transmittance of 0.

Figure 8.10.

COLOR PHOTOGRAPHY

All color films consist of three layers containing silver halide (silver iodide or silver bromide) emulsion, coated on top of each other to form a permanent, multi-layered structure supported by the film base. Each emulsion layer of color negative film absorbs a specific color, determined by the properties of the dyes embedded in that layer. There is a chemical reaction in the emulsion that permanently modifies the layer after it is exposed to and has absorbed light of a specific color. The top emulsion layer absorbs B light and transmits G and R light to the lower layers. The middle emulsion layer responds only to G and the lower responds only to R.

Objects with colors other than B, G, or R, which illuminate the film during exposure, react with two or more of the emulsion layers, according to the spectral distribution of the light they scatter. The original color of the object is reproduced only when the negative is printed. *Color negative film* produces an image that is complementary in respect to colors and density of the original objects. Subsequently, the negative is printed on color photographic paper to yield *a positive color print*.

After exposure, the negative film is developed to yield the color negative. The processing steps include: immersion in a color developer, a bleach rinse, a fixer, a bleach/fix, a final wash, a stabilizer, and drying. Let us briefly look at these steps separately.

The *chromogenic developer* converts light-sensitive crystallites of silver halide compounds in the emulsion layers into metallic silver. While doing this, the developer also oxidizes and combines with dye couplers that are either built into the emulsion layers or added during development. The result is the formation of three dye layers, one from each of the subtractive primary colors: cyan, magenta, and yellow. The blue-sensitive layer of the original film forms a yellow image, while the green and red-sensitive layers form magenta and cyan images, respectively. *Bleach* is then utilized to remove all of the silver metal so that each layer is left with only a color image. After bleaching, the film is *fixed* and *washed* thoroughly to remove all dissolved silver salts. Then it is rinsed in *stabilizer* to improve dye stability and to harden the emulsion. Finally, it is dried.

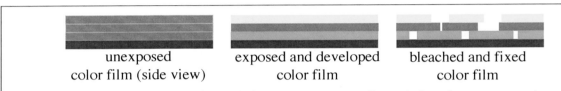

| unexposed | exposed and developed | bleached and fixed |
| color film (side view) | color film | color film |

Figure 8.11. The unexposed film (left) appears gray. Once it has been exposed and developed the B, G, and R absorbing layers are colored. After bleaching and fixing, the silver halide crystals are removed and only the color layers remain, but only in the appropriate places to produce a negative of the image. Where only the cyan layer is present, the image will be printed red, and so on.

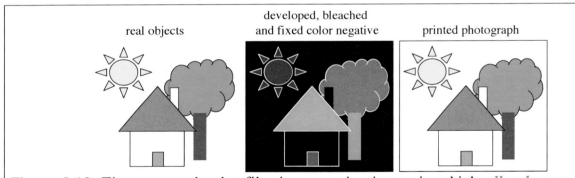

| real objects | developed, bleached and fixed color negative | printed photograph |

Figure 8.12. The processed color film is a negative image in which *all colors are inverted*—that is, the colors are complementary to colors of the original objects. These colors, e.g. the cyan roof and the blue sun, can be seen in transmission on the negative and act exactly like the combination of subsequent CMY filters. In the negative, the C layer is only present where the roof is located in the image and the M and C layers are only present where the sun is. The Y layer is absent from both areas of the processed negative. The black areas on the negative have all three layers. White light is then transmitted through the processed film and projected onto color sensitive paper for printing. The paper is coated with a similar emulsion that inverts the colors all over again!

PIGMENTS

The color of familiar objects depends on the *pigments* that they contain and the *reflectance R* of these pigments. Consider the leaf of a tree, illuminated by sunshine. Some of the light from the sun is specularly reflected as unchanged white light by the glossy surface of the leaf. The remainder of the light is partly absorbed and partly scattered in all directions, as shown in Figure 8.13.

The reflectance is the fraction of the incident light that is scattered. As in the case of transmittance, the reflectance usually depends on the wavelength of light.

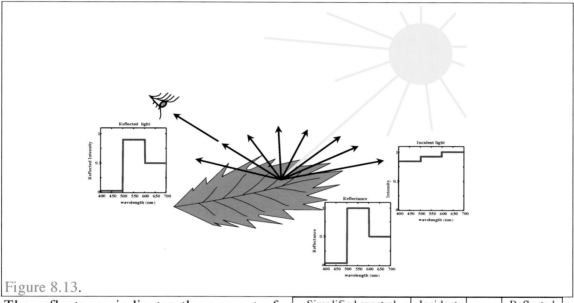

Figure 8.13.

The reflectance indicates the amount of light scattered by the leaf in each of the separate RGB ranges, as shown in the table on the right. The reflectance determines the color of the leaf, as we see it.

Simplified spectral ranges	Incident intensity	R_{leaf}	Reflected light
B (400-500 nm)	0.8	0	0
G (500-600 nm)	0.9	1	0.9
R (600-700 nm)	1	0.5	0.5

Pigments are the colored particles suspended in paints, inks, color photographic film, etc. When two paints of different colors are mixed, a third color is obtained, which is different from each of the two components. Can we anticipate what color will result? Artists do this regularly, using their "experience", that is, after preparing and seeing many mixtures of different colors, they can "remember" the resulting colors. Here we analyze the physical reason for obtaining the color of the mixture. Pigments act very nearly as filters, and are mostly subtractive. The reason pigments in a mixture "subtract" color from each other is called *multiple scattering*. Let us explain this concept with an example: the mixture of red and blue paints. Red pigments absorb all blue and green from the illuminating white light and diffusely reflect only red light. The reflected red light is in turn almost completely absorbed by nearby blue pigments. A similar but inverse behavior is observed at all blue pigments. Red and blue pigments, therefore, absorb each other's scattered light. This phenomenon happens not only once, but many times in the paint mixture, that is why we said the phenomenon is called *multiple* scattering. There is a simple experiment that proved how *multiple* this multiple scattering is. Adding a small amount of black to paint of any other color has a large effect (darkening) on the color of the mixture. See how striking this effect is, in the experimental data of Figure 8.14.

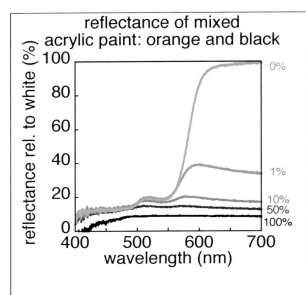

Figure 8.14. The reflectance curves of orange acrylic paint, black acrylic paint, and mixtures of the two, in different proportions. The mixed volumes were measured using Gilson pipettes and the percentage of black paint is reported for each curve by the right axis. The colors used to plot each spectrum are representative of the color of paint or paint mixture. Data from Gilbert and Haeberli, American Journal of Physics, submitted.

Quantitatively, when mixing pigments of two different colors, the resulting color depends on the reflectance curves for the two colors. The reflectance values in each spectral range can be calculated as described above for the transmittances of filters.

It is worth noticing that the reflectance values in each spectral range correspond exactly to the RGB intensities we discussed in additive color mixing, and all colors, when defined *only* by their reflectance or transmittance curves, can be located on the color triangle. All you have to do is use the curves to calculate the fraction of red r and the fraction of green g!

On a computer, magenta and cyan would have the RGB intensities:

M = 255B +0G + 255R
and
C = 255B + 255G + 0R

We already discussed the fact that the units and the absolute values of each color intensity (a, b, and c) are not relevant. The only relevant quantities are the fractions of red and green.

The fraction of red and green for M and C are:

$r_M = 0.5$ and $g_M = 0$;
$r_C = 0$ and $g_C = 0.5$.

Based on these coordinates, it is obvious where M and C are positioned on the color triangle. When subtractively combining colors of known reflectances, if the resulting reflectance is a difficult color to identify, the fractions of red and green can easily be calculated and the color can be identified on the color triangle.

With this new tool, let us now look at a few examples of subtractive color mixing that show changes in saturation or hue.

CHANGE IN SATURATION

Sky B+ Orange = Orange (pure, low intensity)

B=255	B=0	B=0
G=128	G=128	G=64
R=128	R=255	R=128

From the RGB intensities given for each color, we can draw the reflectance curves and fill in the reflectance table.

Simplified spectral ranges	R_{uB}	R_O	$R_{uB\&O} = R_{uB} \times R_O$
B (400-500 nm)	1	0	0
G (500-600 nm)	0.5	0.5	0.25
R (600-700 nm)	0.5	1	0.5

Figure 8.15.

Sky blue is unsaturated B (uB in the table), i.e. B + white. The B reflectance curve therefore has white in it, that is, a certain amount of all 3 spectral components, R, G and B (0.5 in this example), and shown in Figure 8.15. Adding white is equivalent to adding a horizontal line in the reflectance curve that shifts the whole reflectance curve up. Orange, as should be evident from the RGB components and from the reflectance curve of Figure 8.15, is saturated. *As long as at least one of the RGB components is 0, the color is saturated.* As long as the reflectance curve has an intensity of 0 in at least one spectral range, there is no horizontal line added to the curve—there is no white and the color has high saturation.

The reflectance curve for orange is a direct consequence of how orange is additively obtained: 1 part of Y light and 1 part of R light. But Y = R + G lights, therefore orange must contain 2 parts of R and 1 part of G, as shown in the reflectance curve in Figure 8.15.

The color resulting from the mixing of sky blue and orange is saturated, since its reflectivity in the blue range is 0. The reflectance curve for this color is graphed in Figure 8.16. This curve is identical to the one for orange except for a lower intensity: this color is low intensity orange, or brown.

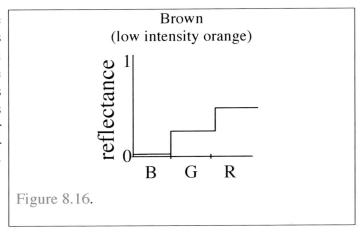

Figure 8.16.

WHY DO BLUE AND YELLOW MAKE GREEN

Have you ever tried to mix blue and yellow paints? If you did, you certainly obtained green. Now, let us look at the idealized reflectance curves for blue and yellow paints, shown in Figure 8.17: there is no spectral region in which the reflectance curves for yellow and blue overlap. In other words, at every wavelength the product of the reflectances is zero. Since this means no light in the 400-700 nm region is reflected and all is absorbed, the color produced should be black. So Figure 8.17 must be wrong. In fact, its spectra are oversimplified. The spectra of light scattered by yellow and blue paints do not have such sharp cutoff, rather they look as the ones in Figure 8.18.

Figure 8.17.

Figure 8.18. The measured reflectance curves for yellow, blue, and yellow+blue acrylic paints. Notice that, since yellow and blue are complementary colors, the resulting mixed color should be black, while in reality is a dark green, with a maximum in the only region in which both spectra are non-zero, that is, green. The color used to plot each spectrum is representative of the color of paint or paint mixture. Data from Gilbert and Haeberli, American Journal of Physics, submitted.

Remember that when we introduced the "oversimplified" spectrum in Chapter 6, composed by only the 3 colors RGB, we said that it works in **most** cases? Well, this is one of the cases in which that simplification does not work. We cannot consider the reflectances of yellow and blue as having 3 values only (0, 1, 1 for yellow and 1, 0, 0 for blue, in the R, G, and B regions, respectively). We must consider that the real reflectance curves have many more values, not only three.

When measuring the reflectance of the resulting paint color, we find that it is non-zero in the green region, as shown in Figure 8.18. That is why the mixture of blue and yellow appears green.

Whenever the simple answer to the question "what color will we obtain if we mix two colors?" is "black", one must be cautious, and not trust the oversimplified spectrum with 3 colors only. There may, in these cases, still be significant overlap of the real reflectance curves of the two colors. In all other cases of color mixing we discussed before and will discuss later, the oversimplification of considering only 3 colors (RGB) in each reflectance curve does not constitute a problem and the results can be trusted.

CHANGE IN HUE

We also mentioned that in most cases, mixing pigments or overlapping filters is equivalent. There are a few exceptions to this rule as well. The two examples below only apply to filters. Mixing two parts of the same color with pigments (in paints or inks) will generate twice the amount of the same color and there is no change in hue.

$$\textbf{M-purple} + \textbf{M-purple} = \textbf{Bluish-M}$$

B=255	B=255	B=255
G=0	G=0	G=0
R=128	R=128	R=64

However, the light transmitted by two subsequent, identical magenta-purple filters is not magenta-purple—it is bluish-magenta! Look at the transmittance table for an explanation of this unexpected result, and the reflectance curves for the corresponding colors in Figures 8.19 and 8.20.

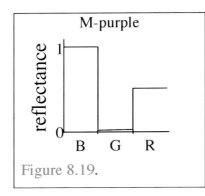

Figure 8.19.

Simpl. ranges	T_{MP}	T_{MP}	$T_{BM} = T_{MP} \times T_{MP}$
B	1	1	1
G	0	0	0
R	0.5	0.5	0.25

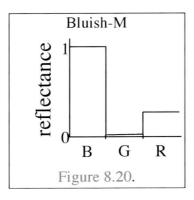

Figure 8.20.

The same phenomenon occurs when using two orange filters, as shown in Figures 8.21 and 8.22. The resulting color is then identified on the color triangle in Figure 8.23.

Two orange filters when used in combination transmit reddish-orange

Figure 8.21.

Simpl. ranges	T_O	T_O	$T_{RO} = T_O \times T_O$
B	0	0	0
G	0.5	0.5	0.25
R	1	1	1

Figure 8.22.

After the first filter

$$r = \frac{1}{1.5} = 0.7 \quad \text{and} \quad g = \frac{0.5}{1.5} = 0.3$$

After the second filter

$$r = \frac{1}{1.25} = 0.8 \quad \text{and} \quad g = \frac{0.25}{1.25} = 0.2$$

Using the values of r and g, we can locate the two hues on the color triangle, as shown in Figure 8.23.

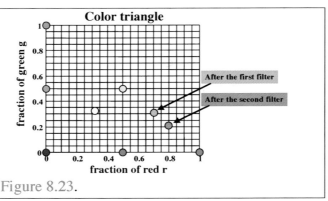

Figure 8.23.

Although these surprising changes in hue do not really apply to mixing pigments, painters are well aware of them. This effect occurs to a limited extent in oil or acrylic painting, but it affects watercolor painting much more significantly. As painters know, one layer of watercolor on white paper is partly transparent and, therefore, acts as a colored filter in front of a projector. If the painter paints a second layer over the first with the exact same watercolor, the hue changes slightly!

Chapter 9
COLOR-GENERATING MECHANISMS

ILLUMINATING LIGHT

The color we see depends on the *illuminating light*. When an apple is illuminated by white light, it appears red, but when illuminated by only blue light it appears black. This is because the pigments in the apple absorb green and blue light and scatter red light. When there is no red light to reflect, all the blue or green light is absorbed and the apple does not reflect light of any wavelength. This is why it appears black.

Colorful pieces of paper illuminated by yellow light, such as a sodium lamp, will scatter all colors that reflect a certain amount of yellow wavelengths (green and red), while absorbing all shades of blue. Any blue shades will therefore appear gray or black.

PIGMENTS

The colors of the familiar objects surrounding us all depend on the pigments they contain. Different pigments absorb and scatter light of different wavelengths, and therefore appear colored themselves. There are animal, plant and mineral natural pigments, as well as synthetic dyes and coloring agents for paints, fabrics, food etc. The animal pigments include:

melanin (which can be black, gray, beige **or brown)**
carotenoids **(which can be** yellow, orange, pink **or red)**

Melanin gives color to our skin and hair, with the exception of red hair, which has a very unusual iron-oxide pigment. On the other hand, carotenoids give color to the beaks and feet of birds, another carotenoid (astaxanthin) colors the shells of crabs, shrimp, prawns and lobsters and becomes the dominant color when these animals are cooked. There are also blue animal pigments, present only in fish and shellfish. The color of blue eyes in humans and Siamese cats is *not* due to pigments, but to scattering.

The most common plant pigments are:

chlorophylls **(which is** green**)**
carotenoids **(which can be** yellow, orange, pink **or red)**
tannin (which is black or brown**)**
anthocyanins **(which can be** red **or bluish-purple)**

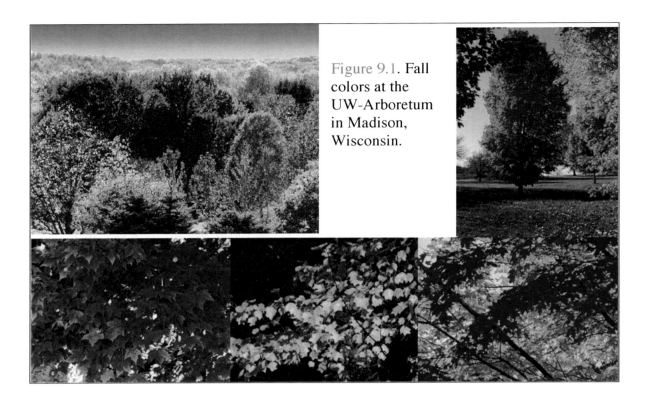

Figure 9.1. Fall colors at the UW-Arboretum in Madison, Wisconsin.

STRUCTURAL COLOR: IRIDESENCE

Some birds, such as parakeets, kingfishers, peacocks, and mallard ducks, have bright blue or green feathers. The shell of many beetles is also green. The wings of the Brazilian butterfly morpho-peleides are a brilliant blue-cyan (Figure 9.2). But with no blue or green pigments in animals, what gives these animals their striking colors?

Figure 9.2. The Brazilian butterfly morpho-peleides, has intense blue-cyan wings.

These colors are generated by *iridescence*. Iridescence is an optical phenomenon separating the spectral colors present in white light. Iridescent colors are generated by the structure of objects, not by the presence of colored pigments. In this sense they are *structural colors*. If a material is subdivided or layered very finely, so finely that the spacing between each subdivisions or layers is similar to the wavelength of visible light, color is generated. Multiple reflections from each layer or subdivision are responsible for this phenomenon, and the subsequent interference of reflected lights determines which colors are seen. See Chapter 1 for interference. The iridescent colors also change with the angle at which the objects are seen.

Iridescent microscopic structure is found on the surface of animals, including: the bird feathers, the beetle shells and the butterfly wings mentioned above. These animal structures are optimized to provoke destructive interference for all wavelengths of white light except green and/or blue. These wavelengths undergo constructive interference, and are then visible to our eyes.

Let us look at the details of iridescence in the wings of the Brazilian butterfly. Light is reflected by microscopic lamellae found on the wings of this butterfly. These lamellae act as tiny mirrors, which reflect light. The size and spacing of the lamellae ensure that only green and blue wavelengths interfere constructively and that red-orange and yellow wavelengths interfere destructively, canceling each other out. The same mechanism explains the color of the peacock, parakeet, kingfisher, and mallard feathers, as well as the beetle shells.

If you were to alter or destroy the microscopic structure of these lamellae, for example by grinding the butterfly wings or bird feathers very finely, the color would disappear since there is no blue-green pigment present.

MORE COLOR-GENERATING MECHANISMS DUE TO IRIDESCENCE

Iridescence colors can also be seen in soap bubbles, thin layers of oil on water, compact discs, pearls, opals, etc.

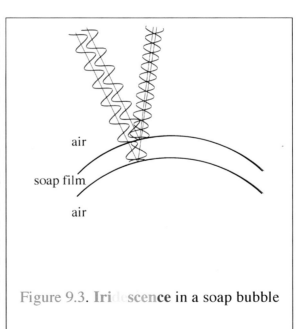

There are only two surfaces generating iridescence in a soap bubble, so this is the simplest of the iridescent systems. Light waves are reflected by both the outer and inner surfaces of the soap film that forms the bubble. The light wave reflected by the outer surface interferes with the one reflected by the inner surface. If these two waves are counter-phase, as in Figure 9.3, they cancel each other out because of destructive interference. If they are in phase, however, they interfere constructively. Consider white light coming from a specific angle and a soap bubble whose film thickness is such that only green wavelengths cancel. In this case, both red and blue wavelengths are reflected and then add together, causing the surface of that portion of the soap bubble to appear magenta.

Figure 9.3. Iridescence in a soap bubble

The thickness of the soap bubble is constantly changing: the top becomes thinner and thinner until the bubble bursts and the water-soap mixture also moves because of convection. Furthermore, the surface of a bubble is curved, and therefore the film thickness through which light travels varies with the angle of incidence. All these factors combine to guarantee that all colors of the rainbow will appear on a bubble's surface, as shown in Figure 9.4.

The same effects take place on thin layers of oil floating on water, as shown in Figure 9.5.

Figure 9.4. Part of a soap bubble with enhanced iridescent colors. Notice that magenta is *not* a spectral color: in this photograph it results from the additive combination of red and blue light.

Figure 9.5. A thin layer of oil on water

In Figures 9.6 and 9.7 we show the colors generated by iridescence in a mineral and a biomineral, opals and mother-of-pearl, respectively.

Figure 9.6. Mother-of-pearl (or nacre) in the inner side of the abalone shell. Iridescence is generated by the regularly repeating, 500-nm thick, layers of calcium carbonate tablets.

Figure 9.7. Australian boulder opals, formed by micro-spheres of amorphous silicon oxide. These are arranged regularly into gratings, and spaced at distances between 400 and 800 nm (similar to the wavelength of light).

Light reflected by the mother-of-pearl layers or the opal silica gratings interferes destructively or constructively for different wavelengths, generating the colors seen on the shell and the opals in Figures 9.6 and 9.7. Interestingly, a rotation of the shell or the opal in front of the observer makes the same locations change color. This phenomenon is also called "opalescence", and is due to the change in effective spacing of the opal grating when this is rotated. Pearls (Figure 9.8) are also made of nacre, as abalone, and their micro-structure is regularly layered, so that each semi-transparent calcium carbonate layer is reflecting part of the light and transmitting the rest to lower layers.

Figure 9.8. Tahitian pearls and Japanese white pearls.

Interference of the light reflected by subsequent layers generates the iridescence, which in pearls is called "pearlescence". Although Japanese pearls are white and Tahitian pearls are dark gray due to different composition, they both have extensive iridescence. Therefore, in both variations you can see a wealth of changing colors as you move them. Pearls are produced by a variety of organisms, most commonly sea mollusks (such as pearl oysters). They are also produced by fresh water mussels and, occasionally, by snails. Pearls are produced as crystals of aragonite are deposited around an irritant such as a grain of sand, a piece of shell, or a parasite. Only the pearls from mollusks with a nacreous mother with pearl lining are "true" pearls. Most pearls on the market are *cultured*—that is, produced by inserting a marble bead into the body a pearl-producing mollusk. The mollusk treats the bead as an irritant and deposits a nacreous coating over it through a process that takes approximately 2 years.

Figure 9.9. Iridescent clouds, gleaned from the website http://www.psi.edu/iridescent.html.

On rare occasions, iridescence can also be seen in clouds. When this happens, clouds are formed by a myriad droplets of water, all having the same size of 1 micron (1000 nm). In these conditions the droplets act as in the rainbow (see Chapter 2), but the clouds display much more intense colors than rainbows against clear skies. Figure 9.9 shows stunningly colorful iridescent clouds.

COLOR IN GEMSTONES

There are several mechanisms that generate color in minerals. Besides the obvious ones based on absorption of certain wavelengths and scattering of others that we have already discussed, the presence of single-atom impurities, their concentration, the specific sites in the crystalline structure of a mineral in which these impurities reside, the oxidation state of the impurities, etc. For example, pure *corundum* is an oxide of aluminum (Al_2O_3). It is a colorless mineral. However, if corundum contains even a small amount of chromium (Cr), it becomes pink or red (and is then called *ruby*). In this case, the impurity is the presence of Cr. Chromium impurities, however, cause *ruby* to be red, *emerald* to be green, and most interestingly *alexandrite* to assume different colors (purple, red, green, or blue) depending on the illuminating light!

Figure 9.10 shows all these gemstones, differently colored by the same element, chromium.

Figure 9.10. Ruby (top left), emerald (top right) and alexandrite illuminated by fluorescent light (bottom left) and candle light (bottom right).
Rough emerald is the author's photos, Ruby and Alexandrite photos are courtesy of Jillian F. Banfield, from her course web site: http://socrates.berkeley.edu/~eps2/.

Different levels of impurity (1-2%) in the same mineral (e.g. beryl, which is $Be_3Al_2Si_6O_{18}$) generate different colors. Here are some examples of beryl gemstones, with different impurities:

BERYL
Aquamarine = blue beryl: Fe^{2+}, this beryl is cyan
Heliodor = golden beryl : Fe^{3+}, this is yellow

Green beryl : mixtures of Fe^{2+} and Fe^{3+} is green
Morganite : Mn^{2+} is pink
Red beryl : Mn^{3+} is red
Emerald: Cr^{3+} is ... emerald green!

Figure 9.11 shows all the different colors of beryl gemstones.

Figure 9.11. Beryl gemstones have different colors due to different impurity atoms (Fe, Mn or Cr) and their oxidation states. Photos courtesy of Jillian F. Banfield, from her course web site: http://socrates.berkeley.edu/~eps2/.

The *alexandrite effect* (a color change as the result of a different type of illuminating light, as shown in the figure above) is an excellent example of how transmission and absorption of specific wavelengths determines color. Light from a light bulb on a dimmer or a lit candle is rich in red and yellow wavelengths while fluorescent light is rich in blue and green. Alexandrite can transmit and reflect blue, green, and red light. Therefore, if the light illuminating alexandrite is rich in red and blue, it will transmit both red and blue and appear purple. On the other hand, if the illuminating light is rich in red but poor in blue and green (candle light), this same piece of alexandrite will look red. If the light is rich in blue and green but poor in red (fluorescent lights), it will look blue or green.

MINERAL COLOR DUE TO CHARGE TRANSFER

Another mechanism generating color in gemstones is charge transfer. Charge transfer involves the exchange of an electron between two atoms. For example blue sapphire (Figure 9.12) is colored by iron (Fe) and titanium (Ti), and it appears blue because R and G light wavelengths are absorbed to transfer electrons back and forth between Fe and Ti atoms.

Figure 9.12. Blue saffire, which is corundum (Al_2O_3) with Fe^{2+} and Ti^{4+} atoms exchanging an electron to become Fe^{3+} and Ti^{3+}, and vice versa. Photo courtesy of Jillian F. Banfield, from her course web site: http://socrates.berkeley.edu/~eps2/.

MINERAL COLOR DUE TO COLOR CENTERS

Color centers are very small (of an atomic-scale) defects in the crystal structure of the gemstone. Even a single electron, when trapped within the crystal structure of a gemstone, is a color center. The electron absorbs some wavelengths of light and uses this energy to "jiggle" within its trap. The remaining wavelengths are scattered, generating the color of the gemstone.

Since these color centers are defects of the crystal structure, they can be "healed" by heating the gemstone or exposing it to UV light. Heat allows the crystal to reorganize itself, while UV light frees the electrons trapped in color centers. Both processes result in removal of the color center, causing the gemstone to change color. If the only source of color was the color center, the stone will become colorless.

Exposure of gemstones to high-energy particles produces color centers. High-energy particles include:

- accelerated electrons
- accelerated alpha particles
- gamma ray photons from a radioactive cobalt-60 source
- neutrons from a nuclear reactor

Color centers give color to naturally yellow topaz. White topaz can also be irradiated and heat-treated to make it blue (London blue, Swiss blue or sky blue), brown or green. Figure 9.13 shows Swiss blue topaz. After irradiation the gemstones are radioactive, and cannot be used until the radioactivity has decayed to safe levels. This may take months or years, depending on the country of origin of the topaz, the irradiation energy and the exposure time.

Figure 9.13. Swiss blue topaz, obtained by irradiation with either neutrons or high-energy electrons and subsequent heat-treatment of natural white topaz.

COLOR IN GEMS DUE TO BAND GAP ABSORPTION OF LIGHT

In materials such as "semiconductors" and "metals", some electrons are no longer associated with individual atoms, but are free to roam around the entire crystal. Just like in pigments and in color centers, electrons in these materials can absorb light at certain wavelengths, thereby generating color in the light they scatter. In semiconductors, however, there is an additional phenomenon: light is only absorbed when its wavelength is below a threshold value. Recall that smaller wavelength corresponds to a greater energy (see Figure 1.6), and for light to be absorbed by a semiconductor it must have enough energy to jump the electrons over a forbidden energy gap, called the *band gap*. Light with too large a wavelength, therefore, is transmitted and the material appears transparent. Materials that have a band gap at wavelengths greater than the visible range (in the infrared) appear completely opaque to us (e.g. silicon), those that have the band gap below the visible range (in the ultraviolet) are transparent (e.g. diamond).

If the band gap falls within the visible range, some colors will be absorbed and others transmitted and/or scattered, therefore the material appears colored. The color in this case may be yellow, orange or red, depending on where the threshold wavelength lies in the optical spectrum, but not blue. Because the transmittance and reflectance are never large at low wavelengths and low at high wavelengths, blue is impossible in pure semiconductors.

The presence of impurities, however, may introduce additional electron energy states within the forbidden band gap. For instance, impurities in diamonds provoke band gap states that absorb specific wavelengths and generate color. In diamonds, the presence of nitrogen impurities generates a yellow color, while boron generates a blue color. Figures 9.14 and 9.15 show diamonds with a variety of colors.

Figure 9.14. Diamonds of multiple colors: yellow and brown diamonds contain nitrogen impurities in the band gap. Photo from Gemological Institute of America website.

Figure 9.15. The famous Hope Diamond, valued more than $200 million, is a large (45.5 ct.) blue diamond, made blue by boron impurities in the band gap. It is surrounded by other colorless diamonds, and exhibited at the National Museum of Natural History in Washington DC.

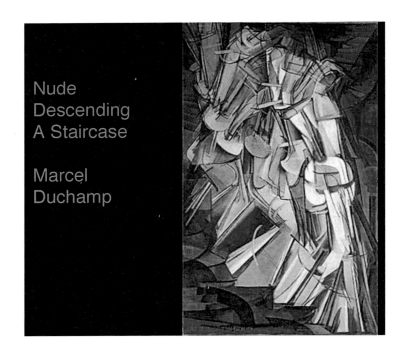

Nude
Descending
A Staircase

Marcel
Duchamp

Chapter 10

PERIODIC OSCILLATIONS

Why do we hear something when we tap a tuning fork, as the one in Figure 10.1, with a rubber hammer? Let's start with a rough explanation of the sequence of events:

1. The two ends of the fork (tines) swing toward and away from each other rapidly - the motion is too small and too rapid to be followed by eye, but by touching the tuning fork lightly you can feel the vibrations. For the "standard A" tuning fork, used by the concertmaster for tuning the orchestra, there are 440 to and fro oscillations every second.

2. As the tines of the tuning fork swing back and forth, air near the tuning fork is pushed back and forth. This causes a slight variation in air pressure near the tuning fork, in step with the motion of the fork.

3. The pressure variations spread out in the surrounding air, much like the ripples on the surface of a lake spread out when you toss a rock in the water.

4. When the pressure variations reach your ear, the eardrum is moved, ever so slightly, in step with the pressure variations.

5. The motion of the eardrum is converted to nerve pulses, which your brain interprets as the sound of a tuning fork. Your brain does not register each oscillation individually. Rather, the oscillations are encoded in the ear in an interesting way, which we will discuss in Chapter 16.

Figure 10.1.
The tines of the tuning fork vibrate and produce sound.

In order to understand sound – how it is produced, transmitted through the air, and perceived - one must begin by understanding mechanical oscillatory motion. Let us first learn what makes objects oscillate.

Vibrations, or oscillatory motions, are not unique to sound. Oscillations are quite common in nature. Familiar examples are slow mechanical oscillations: a swinging pendulum, a weight bouncing up and down at the end of a spring, a squirrel swinging up and down after jumping on a tree branch ,or a plucked guitar string In all oscillations there is a "to and fro motion". What else is common to all oscillations?

1. *A position of equilibrium*: a place where the oscillating object would remain at rest if it were left alone. In the case of the pendulum, the equilibrium position is at the bottom of its swing.

2. *A restoring force:* a force that always pulls the object back to its point of equilibrium once it is displaced from its equilibrium position. If you pull a person on a swing away from their point of equilibrium, gravity will pull them back.

3. *Inertia*: when a mass is in motion, it keeps going (think e.g. of a spacecraft in outer space) unless there is a push or pull to change the velocity of motion. We say that the mass has "inertia", i.e. the mass resists a change in its state of motion. Therefore, if an object is stationary, it will stay put. If it is moving, it will keep moving until another force stops it. Without inertia, a person on a swing would not keep moving back and forth, but simply return to the point of equilibrium and go no further - swings wouldn't be much fun.

As a simple example, let us look at a time sequence of the motion of an oscillating object, such as the one in Figure 10.2.

Figure 10.2. The position x of a pendulum at different times t

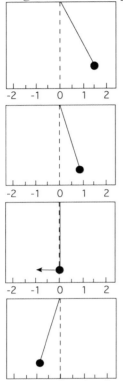

t = 0 sec, x = +1.5m
Somebody pulled the pendulum to the right and released it. If we count the displacement to the right of the equilibrium position as positive, the initial displacement shown in the picture on the left is x= +1.5m.

t=1 sec, x=+1m
Gravity applies a downward force on the mass, but the string keeps the mass from falling straight down. Instead, the mass is pulled back toward the lowest point (equilibrium position or rest position). The pull of gravity speeds up the mass. We say that the *force* acting on the mass *accelerates* the mass.

t = 2 sec, x = 0
The pendulum is at the equilibrium position but keeps moving because of inertia - there is no braking force to stop it.

t = 3 sec, x = -1 m
Now the pendulum is being slowed down by the force of gravity, which pulls the pendulum to the <u>right</u>, toward the equilibrium position, but the pendulum still keeps going until.....

t = 4 sec, x = -1.5 m
....the continuous downward pull of gravity stops the pendulum at −1.5 m. But gravity again pulls it back towards the equilibrium position, and so on. You can complete the sequence on your own.

A more quantitative statement about motion is expressed by Newton's law:

$$\boxed{\vec{F} = M \cdot \vec{a}} \quad \textbf{(10.1)}$$

In words: *force equals mass times acceleration.* Acceleration <u>*a*</u> means how much the velocity changes per second, and force means the amount of push or pull on the mass M.
Since acceleration means the amount of velocity change in *one* second, after t seconds the change in velocity would be:

$$\boxed{\Delta \vec{v} = \vec{a} \cdot t} \quad \textbf{(10.2)}$$

change in velocity = acceleration times how many seconds (t) the force acted.

The Greek symbol Δ (delta) is commonly used to denote a *change* in a quantity. The arrow over the symbols means that the quantity has a particular *direction,* i.e. if the force is in the downward direction then the change in velocity too is in the downward direction. Note that the velocity might well be upward at a given instance in time (e.g. a ball thrown vertically upward), but gravity (a downward force on the ball) would tend to pull the ball downward, and the velocity will therefore decrease. The upward velocity decreases, it is less and less every second, until the ball stops and starts falling back down.

The acceleration of gravity on earth is 9.8 m/s². This means that for a falling object, after e.g. 2 seconds the pull of gravity adds an additional velocity of $(2s) \times (9.8 \text{ m/s}^2) = 19.6$ m/s in the downward direction.

Exercise:
use equation 10.1 to find the force of gravity on a 5 kilogram bag of apples.
Answer:
Mass = 5 kg, a = acceleration of gravity = 9.8 m/s². F = 5×9.8 kg m/s² =49 kg m/s².
For brevity, 1 kg m/s² is called 1N (pronounced "one Newton"). So the answer is <u>49 N</u>.

Example: If you throw a ball up vertically with a velocity of 30 m/s, what will the vertical velocity be after two seconds? Well, you had 30 m/s upward to start with and you added $(2s) \times (9.8 \text{ m/s}^2) = 19.6$ m/s in the downward direction, thus there will be left (30–19.6) = 10.4 m/s in the upward direction.

DISPLACEMENT GRAPH: position x changes with time t

Let us make a graph showing the position x of the pendulum in Figure 10.2 at various times t: we will plot the *displacement x* i.e. distance from equilibrium position (in meters) versus *time t* in seconds (s). We choose to call displacement to the right of the equilibrium position positive, and to the left, negative.

We are going to simplify matters, as shown in Figure 10.3, and only think about the motion of the pendulum in the x-direction (parallel to the ground).

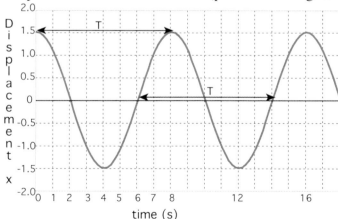

Figure 10.3. The graph shows how the displacement x changes with time. Here we plotted time on the horizontal axis of the graph, and displacement x vertically.

Note that a graph is an abstraction: the actual motion of the object may be horizontal, but that does not keep us from plotting the displacement on the vertical axis.

Looking at the graph in Figure 10.3, one can make several observations:

1. The pendulum swings at most 1.5 m to either side of the rest position. This distance is called the *amplitude A* of the oscillation, just like in Figure 1.4 for light waves. Here A = 1.5 m.

2. The motion we are looking at here is called *periodic*, because it repeats itself over and over again. The time taken for one complete back-and-forth oscillation is called the *period T*. We see from Figure 10.3 that the period in this example is T = 8 s, where s (or sec) is used as an abbreviation for "seconds". Note that it does not matter where one starts counting the time: T is marked in 2 different but completely equivalent ways in the graph.

3. Instead of referring to the period T one often speaks of the number of oscillations (or cycles) per second. The number of complete back-and-forth oscillations per second is called the *frequency f*. The unit of frequency is called *Hertz*, abbreviated *Hz* (named after a German physicist who studied high-frequency oscillations, which were later used for radio transmission). So instead of saying that a tuning fork has a frequency of 440 oscillations per second, we say that the frequency is 440 Hz (pronounced "440 hertz"). Years ago, the abbreviation "cps" was used for "cycles per second" but the current international convention is to use Hz.

4. Different wave shapes: if we plot the motion of a tuning fork, we see a graph very much like Figure 10.3, except that the time scale is much faster. The simple wave shape shown in Figure 10.3 is referred to as a *sinusoidal motion* or as "simple harmonic" motion – what exactly this means we will come to in the next chapter. The tone produced by simple harmonic motion of the tuning fork in music is called a *pure tone*. Most musical instruments (and the voice) make more complicated oscillations when they play a steady tone – the oscillations are periodic but the shape of the pressure variation vs. time is different for each instrument. Two examples of *periodic* but *non-sinusoidal* oscillations are also shown in Figure 10.4.

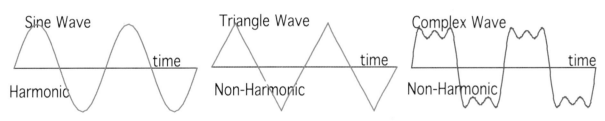

Figure 10.4

THE PERIOD T AND THE FREQUENCY f ARE RELATED:

As you may have imagined already, the period and frequency are related quantities. If the period is very short, there are lots of oscillations in one second and thus the frequency is high. If there are f oscillations in one second, then the time T required for each oscillation must be 1/f seconds. Mathematically:

$$\boxed{T = 1/f} \quad \text{or} \quad \boxed{f = 1/T} \quad \textbf{(10.3)}$$

Anything that oscillates with a *high* frequency f must have a *short* period T, and vice versa.

Example: The "standard A" tuning fork that is used in tuning instruments has a frequency of 440 Hz. Find the period of this tuning fork.

Answer: $T = \dfrac{1}{440\text{Hz}} = \dfrac{1}{440(1/s)} = \dfrac{1s}{440} = 0.00227s$

LARGE AND SMALL NUMBERS: giga-, mega-, kilo-, milli-, micro-, nano-.

In the above example you already noticed that the period turns out to be a small number with several zeroes after the decimal point. The numbers get even smaller when the frequency is high, such as a dog whistle, which may have a frequency of 30,000 Hz.

It is convenient to express large and small numbers by the use of pre-fixes. Some of these you already encountered in the chapters on light, such as nanometers (nm) when you specified wave lengths.

	pre-fix	such as in	Example
1000 = kilo-	k	kg (kilogram), kHz (kilohertz)	30,000Hz = 30kHz
1,000,000 = mega	M	MHz (megahertz) – FM station	88,700,000 Hz =88.7 MHz
10^9=giga	G	GHz (cordless phone)	2.7 GHz=2,700,000,000Hz
1/100 - centi	c	cm (centi-meter)	person height 1.7m=170cm
1/1000 = milli-	m	ms (millisec), mg (milligram)	0.00227s = 2.27ms 0.03g medication = 30mg
1/1,000,000 = 10^{-6} = micro	μ	μs (microsecond) μm (micrometer)	diameter of atom = 0.01μm = 0.000,000,01 m
1/1,000,000,000 = 10^{-9} = nano	n	nm (nanometer)	Wave length of red light = 0.000,000,7m = 700 nm

SPEED OF MOTION

By looking at the graph (Figure 10.3) one can deduce how the object speeds up and slows down. For instance at t = 1.5 s, the displacement x is about +0.5 m, and a half second later (t = 2s) the position is x = 0. Thus the average speed during this time interval is v = (distance covered)/(time used) = 0.5 m/0.5 sec = 1.0 m/sec. Here we don't care about the direction of motion - we simply ask how far the object moved in 1s. We are talking about the *speed* of the object, while *velocity* also specifies the direction of motion. If we count velocity as positive in the +x direction, it would then be negative in the –x direction. If a car goes into a curve at constant speed (using cruise control) does the speed change? No, it doesn't. Does the velocity change? Yes, the direction of the motion is changing.

What we have used here is the concept of speed as you know it from driving a car: the speedometer tells you the distance covered per hour if you were to keep going at the same rate for the entire hour. What the speedometer reads is what we call the instantaneous speed (the speed at that instant) – even though it is labeled as m iles per hour (mph), it does not imply that we keep going for an entire hour. Mathematically we express the speed as

$$speed = \frac{distance}{time} = \frac{\Delta x}{\Delta t} \qquad \textbf{(10.4)}$$

where Δt means a small time interval, and Δx is the distance covered during this time interval.

You might well say that science is not supposed to use wishy-washy concepts like "small time interval" because how would we decide if a time interval is "small"? "Small" compared to what? Well, the idea is that the time interval is small enough that the motion can be considered steady during this time interval. So the required smallness depends on whether the motion is very smooth, changing slowly in time, or whether it is very abrupt and changing rapidly. If you look at the graph in Figure 10.5, the time interval for which we figure the speed should be short enough that the curve during this short interval is a straight line.

A nicer way to visualize the meaning of the speed at one particular **instant** is to draw a tangent to the graph at that instant: in Figure 10.5 we show the tangent at $t = 9$ s and at $t = 12$ s. At $t = 9$ s the tangent points downward (negative velocity) and has slope of –0.8 m/s, while at $t = 12$ s the tangent is horizontal, which indicates that at that time the mass is at rest for an **instant**.

If you think of the slope of the curve in Figure 10.5 you find that the curve is steepest at $t = 2s$, $t = 6s$, $t= 10s$, i.e. it is steepest when the displacement x is zero. This should make sense to you if you remember that the graph in Figure 10.5 represents the motion of a pendulum. Imagine a swinging pendulum and ask yourself at what position does it move most rapidly. Well, it is fastest at the bottom of the swing, when it passes though the equilibrium position!

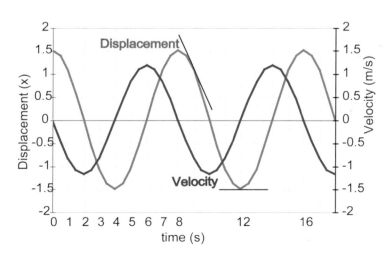

Figure 10.5. The magenta curve, which starts out at +1.5, shows the displacement, i.e. it shows how position changes with time. The blue curve shows how the velocity changes with time. The velocity scale is shown on the right vertical axis of the graph.

Note that the velocity is largest when the pendulum passes through it's rest position (x = 0).

Some of you may have studied calculus – in calculus you learn that as you make the time interval ▢t shorter and shorter (and thus ▢x also gets smaller and smaller) eventually the ratio

$v = \Box x / \Box t$ reaches a limit – strictly speaking it is this limit which is called the instantaneous speed: $v = dx/dt$, i.e. the *instantaneous speed is the derivative of x(t)*. This exactly corresponds to the tangent line we mentioned above. The displacement graph we showed in Figure 10.5 (magenta curve) is a cosine graph – if you take the derivative of the cosine you get a negative sine graph, just as we show in Figure 10.5.

Chapter 11

SIMPLE HARMONIC MOTION

Oscillatory motion requires something elastic or "springy". Deformation of the object causes a restoring force, and this force drives the object back towards the original shape, i.e. toward the equilibrium position. Think of a tree branch: if you pull down on the branch it yields a certain amount, i.e. there is a "displacement". The harder you pull the larger the displacement. A thick, heavy branch will yield less, a thin branch will yield more. Same for a guitar string: when you pluck the string you displace it from the equilibrium position. When you release the string it snaps back – there is a restoring force.

How do we measure the restoring force and how does the restoring force affect the oscillation when you release the branch or the guitar string?

To keep things simple, we use as a springy object a spring or - if you prefer - a rubber band. To find out how springy the spring is, we hang a weight from it and see how much it stretches. A "soft" spring would yield a lot, while a "stiff" spring stretches only a small amount.

Figure 11.1. Different masses are attached to a spring. If each mass is 1 kg, the force of gravity (weight) on each mass is 9.8 N (see exercise in Chapter 10, page 4).

The spring constant k is the elongation x divided by the force (weight) that caused the elongation. If we assume that each added weight of 9.8 N stretches the spring by an additional 5 cm = 0.05 m, the spring constant k is: k = 9.8 N / 0.05 m = 196 N/m.

How do we measure the stretching (called the "*elongation x*")? Simple! We use a ruler and measure how much longer the spring became as we added the weight. Suppose the spring or rubber band was 30 cm long to start with, and after the weight was added it was 35 cm long – the elongation in this case would be (35-30) cm = 5 cm. Usually we measure the length in meters (m), rather than centimeters (cm). Since "centi-" means 1/100, 5 cm = (5/100) m = 0.05 m.

Measuring downward pull (**weight**) of the mass attached to the spring: If you attach a mass to the spring, why does the spring get longer? You will say that the earth pulls down in the mass, and this downward pull is what we call the weight of the mass. Why be so fussy and distinguish between *mass* (how big a chunk you attached to the spring) and *weight* (how much the mass gets pulled down by gravity)? Well, the mass is simply the amount of matter, but the weight depends on where you are: on the moon, the same mass gets pulled down much less than on the earth because the moon is so much smaller, while in outer space, away from all planets, the same mass would be weightless. A conceptually simple way to measure the weight of a mass, be it on earth or on the moon, is to let go of it. The mass will pick up speed during the fall (it will accelerate), but on the moon it will pick up much less speed every second because the pull is less. If you measured the speed after 1 s, 2 s, 3 s, you could find the *acceleration of gravity* (called *g*), meaning how much additional speed the mass picks up every second. As we mentioned in the preceding chapter, on earth, *g = 9.8 m/s²* meaning that every second the speed of a falling object increases by that amount. The units come about from acceleration being a change in speed (m/s) divided by the time (s) so it makes m/s².

To measure the amount of pull (weight), we use Newton's law of motion, *F=Ma* (see Eq. 10.1 in Chapter 10).
If I can measure the mass and the acceleration, I know the pull of gravity, F. Here, the acceleration of gravity is called *g* instead of *a*, so the weight *W* (the force F of gravity on the mass M) is

$$\boxed{\text{Weight of mass M on Earth:} \quad F = M \cdot g} \quad \textbf{(11.1)}$$

Since it may be hard to keep track of all of this, let us simply state the end result:

When you have to figure out the pull of gravity (i.e. the weight) on a mass M, you simply multiply the mass (in kg) by 9.8 and you get the weight in N (Newtons).

Example: If Willy's mass is 75 kg, how large is his weight (on earth)?
Answer: his weight is 75 × 9.8 N = 735 N.

Example: If Willy jumps off a cliff, how large is his (downward) acceleration?
Answer: according to Newton's law, F = Ma. Here F = 735N, M = 75 kg, so
his downward acceleration is: acceleration a = F/M = 735/75 m/s² = 9.8 m/s² – of course, the answer must be just the acceleration of gravity. The acceleration is
the same, for any mass, willy's or any other object or body jumping off a cliff!

THE SPRING CONSTANT

To test springs, we hang different weights from the spring and for each weight we record how much the spring got stretched compared to it's original length. Suppose we increase the weight step by step by adding an additional 0.5 kg mass each time. The results might look something like numbers in the table below:

mass M (kg)	0	0.5	1.0	1.5	2.0
weight (N)	0	4.9	9.8	14.7	19.6
elongation x (cm)	0	1.5	3.0	4.5	6.0

The thing to note is that equal increments in mass (i.e. equal increments in weight) make equal increments in the length of the spring. Or, *the elongation of a spring is proportional to the pull.* The graph of spring length versus force pulling on the spring is a straight line. Of course this is true only up to a point: if you pull too hard on the spring eventually it will be stretched out of shape.
In actual fact there is usually some small measurement error so that the numbers may not be as perfect as in this table.

If F is proportional to elongation x, the results can be expressed as:

$$\boxed{F = k \cdot x} \text{ or } \boxed{k = F / x} \text{ (11.2)}$$

The quantity *k* is called the *spring constant.*

In the above example, for F = 4.9 N pull on the spring the elongation was 1.5 cm. The spring constant in this case is k = 4.9 N/1.5 cm= 3.27 N/cm.
If we convert the elongation to *meters* instead of cm, the answer is k = 4.9/(0.015 m) = 327 N/m.

All other columns of the table would give the same result because F and x increase in proportion.

Not all springy things have a spring *constant*, i.e. not everything stretches evenly in a straight-line fashion when you increase the pull. If you use a rubber band instead of a metal spring you usually find that with equal increments in pull the rubber band yields *more* than it should because the rubber band gets thinner and thinner, making it weaker and weaker, thus easier to stretch.

Stiff springs have a large value of the spring constant, because a stiff spring requires a larger force than a soft spring for a given amount of stretching. Until now, we talked about stretching or elongation of a spring, but the same applies to compression of a spring such as the springs of a car suspension.

OSCILLATION FREQUENCY FOR SIMPLE HARMONIC MOTION (SHM)

Simple Harmonic Motion **is the name given to the motion that results when the force on a mass is exactly proportional to the displacement of the mass from its rest position.**

When the mass hanging from a spring is displaced and then released, it will oscillate. How rapid is this oscillation, what does the oscillation frequency depend on, and what is the wave shape of this oscillation? The answer to these questions will be important later on, when we want to figure out what musical tone is produced, for instance, by the strings of a piano or a guitar.

You are perfectly correct if you guessed that, with some math, one can figure out the frequency of oscillation of a mass on a spring, given that we know the *spring constant k* and the amount of *mass M*. Newton's law, F = ma, says that if we know the spring force F and the mass M, we can figure out the acceleration. Knowing the acceleration, in turn, lets us calculate the velocity at each instant in time, so we can predict how the object moves, based on a simple calculation. But what makes the calculation complicated is that as the mass moves up and down, the position changes, which changes the spring force, which changes the acceleration, which changes the velocity, which changes the position – so we are back to the beginning of the chain of quantities, all of which change with time.

The math required for this calculation is rather elementary calculus. But instead of just telling you the result, let's see if we can figure out at least part of it together. Whenever possible you should try to reason things out rather than just taking somebody's word for it.

When you displace the mass some distance x, there will be a restoring force F pulling the mass back to the equilibrium position. A large spring constant k means the force is large, causing a large acceleration. The mass picks up speed quickly and returns to the equilibrium position in a short time. Conclusion: *large spring constant k causes rapid oscillations (large frequency)*. On the other hand, a large mass is sluggish, picks up speed slowly, so for the same spring as before, *a larger mass will produce a lower frequency of oscillation*. These two statements imply that in the formula for frequency, the spring constant k should be in the numerator, the mass M should be in the denominator – larger k gives larger f, larger M gives smaller f.

So we would think that the vibration frequency depends on k/M. Basically, this is correct, but let us look at the units: the frequency must have units of number of oscillations per second, but does this agree with our answer k/M? If you figure it out (k has units of N/m, N has units kg-m/s², M has units of kg) you find that our answer has units of $1/s^2$, not 1/s. So we need to take the square root of the quantity k/M.

Indeed, the correct answer for the oscillation frequency of a harmonic oscillator is:

$$f = \frac{1}{2\pi}\sqrt{\frac{k}{M}}$$ **(11.3)**

This is called the **natural frequency** of the oscillator, "natural" in the sense that if the mass M is set in motion, and then left to itself without outside interference, f is the frequency at which the mass inherently wants to oscillate. The formula is just what we guessed; the only thing we missed in our guess is the $1/2\pi$ in front. For the formula to work out, we must use the proper units: *k* is expressed in *N/m* and *M* in *kg*. The frequency then has units of 1/s, which means so many complete oscillations per second, or so many Hertz.

Example: At a farmers' market a farmer uses a spring scale to weigh apples. Assume that 2 kg of apples stretch the spring by 10 cm. Find the oscillation frequency of the apples on the spring scale, after they are given a push.

Answer: The weight is 2×9.8 N $= 19.6$ N. The elongation is 10 cm $= 0.1$ m, thus the spring constant is k $=$ F/m $= 19.6$ N / 0.1 m $= 196$ N/m.

The frequency is $f = \dfrac{1}{2\pi}\sqrt{\dfrac{196}{2}}\,Hz = 1.58\,Hz$

Does amplitude affect the frequency? Our first intuitive reaction might be to say that larger amplitude means that the mass has further to travel on each swing so that the period (time for one oscillation) should get longer. Not so! If you make the initial displacement, for instance, twice as large, indeed the mass has twice as far to travel, but at the same time for the larger displacement the spring force is also twice as large. Twice a spring force means twice the acceleration, twice the speed picked up, which exactly offsets the larger distance traveled. So indeed this is the reason that in the formula for the vibration frequency the amplitude does NOT appear! In other words the natural frequency does not depend on the amplitude of the displacement.

But note, frequency does not change with amplitude *only* for a simple spring force, which is exactly proportional to displacement. In all other cases the frequency changes when the amplitude changes. *It is a special property of simple harmonic motion that the frequency of oscillation is independent of amplitude.*

This is of enormous importance in musical instruments. In music, *amplitude* is related to *loudness* of the tone, while *frequency* of oscillation is the *pitch* of the tone. When we say that the frequency is independent of amplitude, we are saying that the pitch does not change when you play softer or louder. The amplitude of a piano string can be large or small, depending on how loud a tone is being played, but the frequency is still the same. If not, it would be a disaster since the musical pitch (which relates to frequency) would change every time the musician goes from soft to loud.

WAVE SHAPE OF SIMPLE HARMONIC MOTION:

To produce a graph (position vs. time) for simple harmonic motion (SHM) we could imagine attaching a pen to the oscillating mass, and sliding a strip of paper at a constant speed under the pen. The result is shown in Figure 11.2.

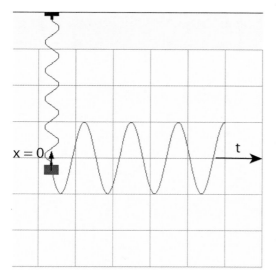

Figure 11.2. Wave shape of Simple Harmonic Motion

A mass hanging from a spring oscillates and makes a trace on a paper strip that moves to the right.

The horizontal axis shows time, the vertical axis is displacement.

It is fun to look at the demonstration on the web: http://www.surendranath.org/Applets/Oscillations /FDHM/FDHMApplet.html

What exactly is the shape of the graph that characterizes simple harmonic motion? Those of you who had some trig will say "just call it a sine curve". Even if you had no trigonometry, in a couple of minutes you can learn that the sine-function is just a name for something very simple, as shown in Figure 11.3: Draw a circle with a radius 1 unit of length, and draw a radial line that is rotated counterclockwise by an angle α ("alpha") with respect to the horizontal. The length of the green line in Figure 11.3 is called the "sine" of angle α, or sin α (read sine alpha). No math skills needed: just a protractor to measure angles and a ruler! In Figure 11.3, the angle is 30° and you see that sin 30° = 0.5. If instead you draw the radius at 45° and count the number of squares you find sin (45°) = 0.707. You also see that the largest value of sinα is found when α = 90°, i.e. sin (90°) = 1.

You might say: why make a graph like Figure 11.3? Didn't it occur to anybody to write a book with values of sine for different angles? Yes, but even more convenient, most pocket calculators have a sin button. If you put in 45 and press "sin" it should show 0.707 on the display. If not, your calculator may use a different way to measure angles. In radians, once around a circle is 2π radians. In this case you have to reset the calculator to use degrees, so once around a circle is 360°.

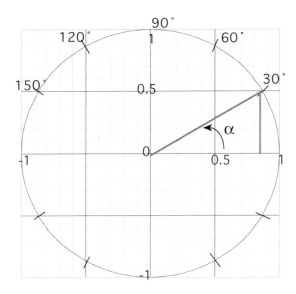

Figure 11.3. Explanation of the sin function.

The circle has a radius R = 1. The red line is a radial line rotated counterclockwise by an angle ▯ with respect to the horizontal.
The length of the vertical green line is called "the sine of the angle ▯". In this example, you find sin (30°) = 0.50. You also note that sin (0°) = 0, and sin 90° = 1. When the angle ▯ is between 180° and 360°, the sine of the angle is negative.

Exercise: draw the radius line for ▯ = 210° in the graph and show that sin (210°)= –0.5. It helps to note that 210° = 180°+30°

An equation to describe SHM: The **displacement x** of a pendulum (or the air pressure for a tuning fork making SHM) can be written as an equation which contains the **amplitude A** of the oscillating object, the **period T** and the **time t** at which you want to know the displacement. The answer is

$$x = A \cdot \sin(360° t / T)$$ or, if angles are in radians $$x = A \cdot \sin(2\pi \cdot t / T)$$ **(11.4)**

Let's see if this makes sense: when one period has passed, the time t equals the period T, and thus t/T = 1. Indeed, after one period the angle in the sin-function must be 360° (or 2▯ if you use radians instead of degrees). The reason to multiply sin by the amplitude A is that the largest value of sin is 1 but the largest displacement is A.

Often we would rather express **x** in terms of the frequency *f* instead of the period **T**. We can simply replace 1/T by f:

$$x = A \cdot \sin(360° f \cdot t)$$ or, if angles are in radians $$x = A \cdot \sin(2\pi f \cdot t)$$ **(11.5)**

Exercise: a tuning fork vibrates with a frequency of 200 Hz and amplitude 1mm.
Make a graph showing the displacement every 0.5ms between time t=0 and time t=5ms. Use a calculator to look up the sin values.
Answer: it is a bit tedious, but easier than you think! Make a table of time values t (0 s; 0.5 ms=0.0005 s; 1.0 ms = 0.001 s; etc). For each value calculate frequency times time. Then get sin of that number from the calculator, and multiply by A - this is the displacement x. The graph should look similar to Figure 11.2.

The displacement graph you calculated in this way is called a *sinusoidal oscillation.*

PHASE ANGLE:

The angle that the rotating arm makes with the horizontal is called the phase angle. When we look at a displacement versus time graph for a harmonic oscillator (a sine curve) we can mark the phase on the curve or on the time axis. The phase angle is particularly useful later when we compare two similar sinusoidal oscillations that are out of step.

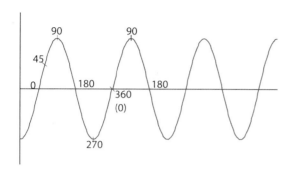

Figure 11.4. A sinusoidal oscillation. The phase angle is marked at different points of the oscillation - it is the angle between the horizontal axis and the radial line in Figure 11.3.

After a complete oscillation, the phase angle is 360° or 0°, whichever you prefer to call it.

Marimba from the highlands of Guatemala

Chapter 12

DAMPED OSCILLATIONS AND RESONANCE

DAMPED OSCILLATIONS - THE CONCEPT OF "DAMPING TIME"

Up to now we have talked about oscillations (pendulum, mass and spring, etc.) as if - once put in motion - these oscillations continued indefinitely. We know that in the real world, oscillations sooner or later come to a stop. The graph that represents displacement versus time does <u>not</u> look like the top one in Figure 12.1, but more like the one at the bottom of the same figure. The amplitude of the oscillation decreases with time. Friction or air resistance is responsible for this behavior. The larger the friction the more rapidly the amplitude of oscillation decreases. The energy that was given initially to the oscillator is being slowly changed into some other form of energy. Some of the energy may go into heat; in a musical instrument (piano) some of the energy may be given off as sound energy.

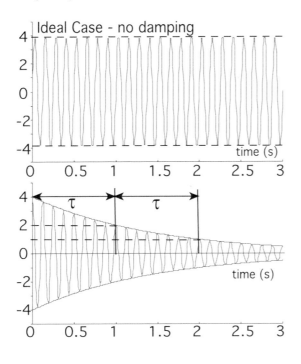

Figure 12.1. Two oscillations of the same frequency are shown. If there is no friction or other loss of energy of the oscillator, the amplitude remains the same. But in reality, there is always some loss of energy that makes the amplitude of the oscillations decrease with time, as shown in the lower graph.

The "damping time" τ is the length of time it takes the amplitude to decease to half of the previous value.

*The time taken for the amplitude of an oscillator to decrease to <u>half</u> its original value is called the **damping time** τ (Greek letter "Tau").* Note that the amplitude does not decrease in a straight-line fashion - rather the decrease in amplitude becomes smaller as the amplitude itself gets smaller. One way to put it is to say that the loss in amplitude is proportional to the amplitude itself. As a result, if the amplitude decreases to half in one damping time τ, it will reduce to a quarter in 2τ, as shown in Figure 12.1. This is called an exponential decrease in amplitude. If you inherit a small fortune and every year spend half of what you still own, your capital would decrease in the same way - you never run out of money! The same with a tuning fork - theoretically it will never stop but after a few seconds the sound is so soft that nobody can hear it. Or you are so poor you can't buy yourself a hot dog!

A long damping time *τ* indicates that there is little friction to dissipate the energy of the oscillator, we say that there is little "damping". A short damping time *τ* indicates that the oscillator is losing energy rapidly - the oscillator is "heavily damped" as in the last graph of Figure 12.2. The vibrations of a drumhead are heavily damped, compared to the oscillations of a plucked guitar string, which lingers for several seconds. A 500 Hz guitar string with a damping time of 2 s makes 2 × 500 = 1000 oscillations before it loses half the amplitude.

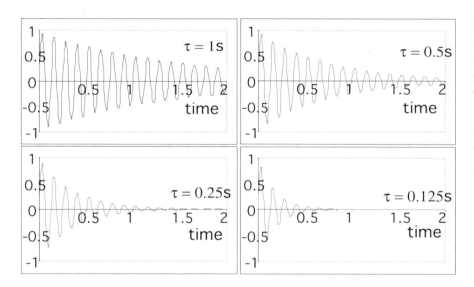

Figure 12.2. Each of the oscillations shown here has a frequency of 8 Hz. The damping time varies from 1 s for the first graph to 0.125 s for the last graph.

<u>RESONANCE</u>

How are oscillations in musical instruments set in motion? In many instruments the oscillations are built up by adding many little pushes, timed to correspond to the period of the oscillator, much like a gentle person can set a swing in motion by many properly timed small pushes. As an example, look at a trumpet. The musician, by pressing his lips against the mouthpiece, makes rapid periodic puffs of air (try to "buzz your lips!) that set in motion the air column in the trumpet. For a trumpet, the build-up of the oscillations is quite rapid but in a large organ pipe, it may take half a second or more to build up the oscillations.
Another example where an oscillating system is set in motion by repeated pushes from the outside is the soundboard of a piano, which is put in motion by the vibrating string.

Here we will learn about the fundamental properties of such "driven oscillators" by first studying simpler systems, such as a pendulum or a cart between springs, riding on a (nearly frictionless) track.

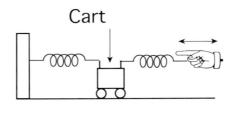

Cart

Figure 12.3. You apply a gentle, periodic force near the top of a pendulum.

Figure 12.4. You apply a gentle, periodic force to a glider on wheels.

We will assume that the external force (the "*driving force*") is a simple *periodic sinusoidal pull* with a constant frequency. Let's consider the example of the cart on wheels in Figure 12.4. There are four things to be taken into consideration:

1. The *natural frequency f_N* (the subscript N is attached to f to indicate that it is the *natural frequency*). *The natural frequency is the frequency of oscillation in the absence of an external driving force.*

   In Chapter 11 we learned that f_N depends on the physical construction of the oscillator, such as the mass of the cart and the stiffness of the springs.
   As an example, let us assume that the oscillator, if it is set in motion and then left alone, has a frequency $f_N = 10$ Hz.

2. The *driving frequency f_D* is the frequency with which you pull on the spring. The driving frequency can be whatever one chooses - you are not obligated to push the oscillator at it's natural frequency.
   For instance you might pull 5 times per second ($f_D = 5$ Hz) or any other frequency.

3. The *amplitude of the driving* force specifies how hard you are pulling on the spring - i.e. how large a force you use. You can choose to pull very softly or very hard; a stronger force will result in larger amplitude of the motion.

4. The *amplitude of the oscillator* measures the response of the oscillator to the driving force.

We want to see how the motion of the glider depends on the driving frequency f_D. We try various frequencies using the <u>same force</u> in all cases, and we *observe the amplitude* of the oscillator. Suppose that the oscillator has a natural frequency of 10 Hz but we start out by pulling on it with a driving frequency $f_D = 2$ Hz. Note that f_D is very different from f_N. The glider starts to move a little bit - it wants to move at 10 Hz, but the pushes are at 2 Hz so the pushes are not at all in step with the natural frequency. In the beginning the motion of the glider is irregular - being neither 10 Hz nor 2 Hz, but it is soon forced into vibrating at the frequency of the driving force f_D. However the amplitude A of the oscillation will be small, as indicated in Figure 12.5.

Now we change f_D to 8 Hz, which is closer to the 10-Hz natural frequency. The first oscillation of the driving push imparts a small amount of motion to the glider - the second back-and-forth push in this cases comes 0.125 s after the first push, but this is very close to the 0.1 s that would be needed to be in step with the glider motion. This means that for f_D = 8 Hz, the glider moves with a larger amplitude than for f_D = 2 Hz where the timing of the pushes was completely off.

To build up the biggest amplitude, the driving frequency f_D should be equal to f_N because then the subsequent pushes keep adding to the motion imparted by the previous pushes. So at f_D= 10 Hz the cart's amplitude will be the largest. When you push with too high a frequency, the amplitude is less again because the pushes are out of step with the natural frequency.

A plot of the cart's motion amplitude vs. driving frequency is shown in Figure 12.5.

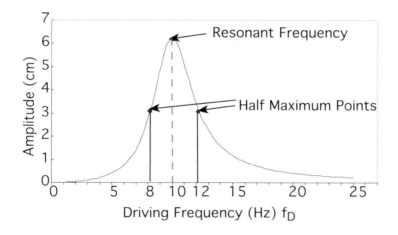

Figure 12.5. Resonance curve for an oscillator with a natural frequency of 10 Hz. In this example, if the driving frequency differs from the natural frequency by 2 Hz either way, the amplitude of the oscillator is still half of the maximum amplitude when you are exactly on resonance. The width of this resonance curve is 2 ☐ 2 Hz = 4 Hz.

The response of an oscillating system to pushes from the outside shown in Figure 12.4 is called "resonance". The oscillating cart resonates at 10 Hz, i.e. it selectively responds to the pushes if they are at the resonance frequency.

For simplicity, we chose to use a pushed cart as an example - what other examples of resonance can you think of? Maybe singing in the shower stall? - if you have a deep loud voice and change the frequency from high to low you may find a big echo at a certain frequency, namely at the natural frequency of the air oscillations in the shower stall volume. A demonstration of resonance you can try, if you have access to a piano, is to depress the loud pedal (which takes the dampers off the strings) and sing a loud tone close to the piano - some string may resonate, namely the strings whose natural frequency is close to the driving frequency of your voice. It may help to hit a key first and then sing the same tone to get your voice's f_D to match the f_N of the string.

How close to the natural frequency (f_N) must the driving frequency f_D be, to give reasonably large amplitudes? One way of looking at this is to find at what frequencies the amplitude is exactly half of that obtained when $f_D = f_N$. In the graph above the maximum amplitude (for $f_D = 10$ Hz) is 6 cm. At $f_D = 8$ Hz the amplitude is half the maximum value. The same is true at 12 Hz. The difference between these two frequencies is called the width (Δf) of the resonance curve (also called the "full width at half maximum" FWHM). In this case $\Delta f = 12 - 8 = 4$ Hz.

Another resonance curve for a different driven oscillator is shown below. In this case we would say the resonance frequency is $f_N = 60$ Hz and the width $\Delta f = 20$ Hz.

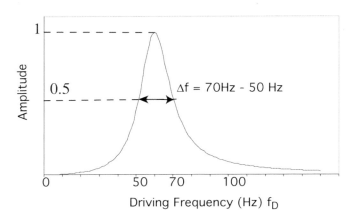

Figure 12.6. Resonance curve of a large Tabla (Indian drum) when set in motion by a periodic force. The resonance frequency f_N is 60 Hz and the width of the resonance is 20 Hz.

A resonance width of 20 Hz means that if the driving frequency f_D is off by half that amount (10 Hz) to either side, the response of the oscillator is half the maximum response.

The width of the resonance curve depends on the amount of damping. One finds that some resonance curves are very sharply peaked, while others are quite broad. The important difference between sharp or broad resonances is the amount of damping (amount of friction) of the oscillating system that is being pushed. If the oscillator has little damping, one can build up a huge amplitude, provided the driving frequency is close to the natural frequency. Why? Because for small friction (long damping times) the oscillator gets a little more energy with each push, i.e. it still remembers all the previous pushes. But in this case, even if the driving frequency is off only a little, the later pushes will eventually have fallen out of step and start to reduce rather than add to the oscillations.

On the other hand, when the oscillator is more damped, the energy it acquired from the pushes is soon dissipated, so the oscillator does not remember the previous pushes. This has two effects, illustrated in Figure 12.7: (a) for more friction, the peak of the resonance curve is not as high, but (b) if you are off in frequency it matters relatively less so that the resonance curve is wider. In summary, there is an inverse relationship between Δf and $\square$: *long damping time (small amount of friction) corresponds to a small width of the resonance curve, and vice versa.*

Above, we always assumed that the pushes of the driving force continued for a long time, i.e. the oscillator came to a "steady state". The resonance curve shows the steady state response to the driving force. The time needed for the oscillations to steady down depends on the damping time: short steadying time for short damping time, long for long damping time.

The inverse relationship between Δf and τ can be put into quantitative terms: a calculation shows that the product of Δf and τ is always the same:

$$\boxed{\Delta f \cdot \tau = \text{constant} = 4/9} \quad \textbf{(12.1)}$$

If you make Δf large, τ becomes small or if you make τ large, Δf becomes small (long decay and sharp resonance). This relationship has important applications and is extensively used in acoustics.

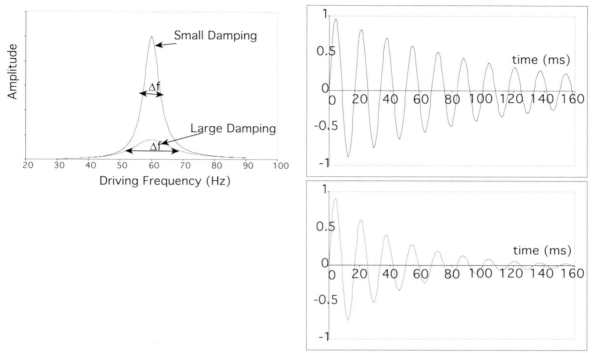

Figure 12.7. The graph on the left shows two resonance curves for different amount of damping (friction), but the same resonance frequency. The graphs at right show the damped oscillations for the same two cases

Exercise: The drum-head of a kettle-drum has a frequency of 50 Hz. The drum-head makes about 10 oscillations before the amplitude has decreased to half the initial value. Find (a) the damping time and find (b) the width of the resonance curve.

Answer:
(a) The period is (1/50) s, so 10 oscillations take (10/50) = 1/5 = 0.2 s Damping time τ = 0.2 s.
(b) Solving equation (12.1) for Δf gives: $\Delta f = 4/(9\tau)$ =(4/1.8s) = 2.2 Hz. Note that the units come out OK: τ is in seconds, so $1/\tau$ has units 1/s which is the correct unit for frequency (Hz).

BUILDUP AND DECAY OF MUSICAL TONES

When you listen to an instrument (e.g. a clarinet or a violin) you get the impression that the sound of a given note played starts and stops instantly. But, if you look at the tone on an oscilloscope, you find that the tone builds up and decays gradually. Build up and decay times are closely related: if it takes the tone 0.1 s to decay to half, it also takes 0.1 s to build up to half the amplitude. It turns out that the gradual build up and decay is an important attribute of the tone quality of an instrument. If one takes the recording of a tone played on an instrument, and with a fast switch cuts off the build-up and decay, it is much harder to tell what instrument is playing!

APPLICATIONS IN MUSIC

Resonators in musical instruments:
Sympathetic strings: some instruments, including the Indian Sitar or the 18th century viola d'amore as well as some guitars, use so-called "sympathetic" strings to enhance the sound. These strings are not bowed or plucked themselves but are set in oscillation by a nearby string.

Instruments with resonators: The marimba and the xylophone have sets of wooden bars of different lengths, producing a sequence of tones similar to the piano keyboard. The sound of the bars is reinforced by hollow tubes (resonators), whose dimensions are matched to the pitch of each bar to produce a resonant sound. The origin of the Marimba is not known, but in the mythology of the Zulus of South Africa, there is a tale about a goddess called "Marimba" who made an instrument by hanging gourds below wooden bars struck by a mallet. A traditional marimba from the highlands of Guatemala is shown next to the chapter title. Another instrument, the xylophone, is similar to the marimba, but has a different musical range. A modern Marimba is shown in Figure 12.8. The frequency of the struck bars increases from one end of the instrument to the other and the resonators correspondingly should increase in length across the instrument. It thus seems puzzling that the bars in front of the instrument shown in the figure are symmetric, short in the center and long at both ends. The answer is that the bars in front of many commercial instruments are ornamental only – the real resonators are hidden behind the ornamental ones, and of course they follow the laws of physics

Figure 12.8. A modern Marimba. The actual resonators are behind the ornamental vertical bars in front.

The *vibraphone* is a modern instrument (often used in jazz)which we mention here because it provides a vivid demonstration of resonance. The instrument has a set of metal bars and metal tubes of different lengths below each bar. Above each resonator is a metal disc, slightly smaller in diameter. These discs are attached to a motor-driven horizontal bar, such that the resonators are periodically covered and uncovered, causing the resonators to be rapidly turned on and off. Correspondingly, the sound intensity is modulated rapidly from loud to soft ("wow-wow-wow"). The musician can change the speed of the motor to change the rate of modulation.

Soundboards of instruments: the sound volume produced by a piano string by itself is very weak because the string itself is small and does not push on much air. The beautiful loud sound of a good piano (or any other string instrument) is achieved by running the taught string over a bridge which transmits the vibrations to a large *sound board*. Some people speak of a resonance board, but this is very poor terminology because the instrument builder wants to avoid resonances. Why? Because a resonance selectively responds to a narrow band of frequencies, while the sound board should respond evenly to all frequencies of the instrument.

A related comment can be made about loudspeakers. Most *loudspeakers* consist of a cardboard cone with a corrugated flexible rim that allows the speaker cone to move in and out. The driving force is provided by an electrical coil and a magnet that pushes on the apex of the speaker in step with the electrical signal from the amplifier. So indeed this is an oscillating system with a periodic driving force. To achieve a more or less uniform frequency response, we want the resonator to be very heavily damped so that the resonance is flat and spread out. To obtain good music reproduction over all frequencies one uses not just one speaker, but separate speakers for bass, ("*woofer*"), midrange and high frequencies ("*tweeter*"). An unavoidable consequence of the need for heavy damping (high friction) is that most of the energy supplied by the amplifier goes into overcoming the friction. A 50-Watt music system produces but a small fraction of 50 Watt as sound energy, most energy produces heat instead of sound.

While not related to music, *touchtone dialing* (technically called dual-tone multi-frequency dialing) is an interesting everyday application of resonance. When you press one of the twelve keys on a telephone keyboard, you hear two pure tones. These tones are generated by your telephone and are sent to the telephone switchboard center. At the switchboard, the signal is sent to seven resonance circuits, each of them responding to one particular frequency. Seven tones are sufficient to represent the 12 keys. For instance, when you push 1, the two frequencies are 697 Hz and 1209 Hz. It is said that some people with good tone perception can tell the number from the tones.

Chapter 13

ADDING SOUND SOURCES: BEATS AND HARMONY

13.1 PRINCIPLE OF SUPERPOSITION

Sound consists of rapid pressure variations. If the pressure variations are periodic, we perceive the sound as a **tone**, and not as **noise**, which consists of disordered (random) pressure variations.

What happens to the pressure variations at your ear (or at a microphone which picks up the sound), when two of more sound sources oscillate at the same time – for instance an entire brass band all playing together?

The way to look at this problem is to say: each instrument, if it played by itself, would produce a certain sound pressure variation with time. Let us call the sound pressure of instrument #1 at a given instant in time $P_1(t)$. Then let us have instrument #2 play and call its sound pressure $P_2(t)$. What is the sound pressure when both instruments play at the same time? The *Principle of Superposition* says that the new pressure is the sum.

Pressure created by many instruments = sum of the pressures created by each instrument:

$$P(t) = P_1(t) + P_2(t) + P_3(t) + ... \qquad \textbf{(13.1)}$$

Example: one flute produces the pressure variations shown as the black curve in Figure 13.1, while the second flute produces the pressure variations shown as the blue curve. When both flutes play, the pressure variations are the sum (bottom graph in Figure 13.1). Note that in adding the pressures we have to take into account that at a certain instant one instrument may produce a negative pressure (pull on the eardrum) while the other produces a positive pressure (push). At the instant marked in the graph by the red arrows (t = 3 ms), P_1 = +1.4, P_2 = -4.0, thus the sum at that moment is P = -2.6. At t = 4 ms both pressures are zero so the sum is zero.

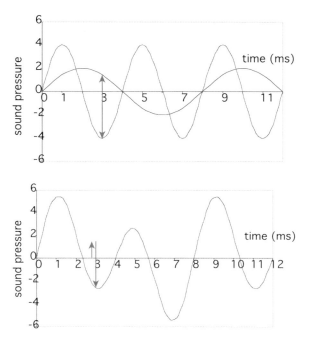

Figure 13.1. Sound pressure variations from two flutes. At time t = 3 ms, one instrument produces a positive pressure of 1.4 units, the other a negative pressure of -4.0 units. The lower graph shows the sum (or *superposition*) of the two sounds.

13.2 THREE SPECIAL CASES

13.2.1. TWO PURE TONES OF THE SAME FREQUENCY

"Pure tones" correspond to sinusoidal pressure variations, e.g. tuning fork oscillations. *If one adds two sine curves of the same frequency, the result is again a sine curve with the same frequency, but the amplitude of the sum depends on the relative timing of the two oscillations.* To understand this, imagine two swings of the same length side by side: they can be set in motion in synchronism (in step), or they might be going out of step with one another. The "out-of-stepness" is measured by a quantity called the *phase shift* ϕ (Greek letter phi), where one full oscillation corresponds to 360° phase shift. Thus, for example 90° phase shift corresponds to 1/4 oscillation, as shown in Figure 13.2. You will note in the figure that the sum is again a sinusoidal oscillation but the amplitude is not simply the sum of the two amplitudes because the two oscillations do not reach the peak at the same time.

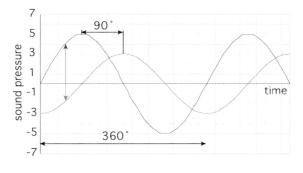

Figure 13.2. Two sinusoidal oscillations of the same frequency but different amplitudes (5 and 3 units of pressure) are added. The oscillations are not in step but are shifted in phase by 1/4 of a full oscillation. If we consider a full oscillation to represent 360°, the phase shift here is 1/4 of 360° or 90°. The figure below shows the sum of the two oscillations.

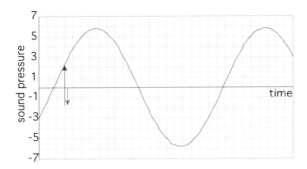

If the two oscillations are shifted by 180°, they completely oppose each other. In this case when one oscillation is at its peak ($P_1 = 5$) the other is at its trough ($P_2 = -3$), and the amplitude of the sum is only 2 pressure units, that is, less than either instrument alone! This is shown in Figure 13.3. Therefore, instead of being louder, the two sounds together are softer than either one alone. Could this really happen? Yes, if you produced the tones electronically and fed them to the same speaker, but not for a musical instrument. Musicians cannot time their instruments so accurately as to remain exactly out of phase for any length of time.

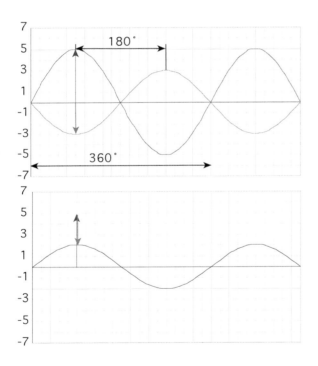

Figure 13.3. Same oscillations as in Figure 13.2, but with 180° phase shift.

How does one measure the phase shift? A simple way is to count how many divisions on the graph you would have to shift one of the curves to get the two curves in phase. Next you count how many divisions one period corresponds to and call this 360° (as in Figure 13.2 and 13.3). Now you use simple proportions to find the phase shift in degrees.

Example: in Figure 13.2 the red oscillation is shifted to the right (delayed in time) by 5 divisions with respect to the blue oscillation. One full oscillation is 20 divisions = 360°. Thus 1 division = 360°/20 = 18°, and 5 divisions = 5 × 18° = 90°.

13.2.2. BEATS

Consider *two tones of nearly the same frequency*, for instance $f_1 = 100$ Hz and $f_2 = 98$ Hz. For simplicity we assume that both tones have the same amplitude. In musical terms, the two musicians play the same tone but are slightly out of tune, as they might be before the performance, when their instruments are not well tuned.

In this case, if the two oscillations are initially in phase (in step) they will slowly go out of phase because one frequency is a little higher than the other, which means that one period is a little longer than the other. In this example, after 1/4 second, the first instrument (100 Hz) has made 25 oscillations while the second instrument (98 Hz) has 98/4 = 24.5 oscillations. This means the corresponding oscillations are shifted with respect to each other by 0.5 oscillations, which corresponds to a phase shift of 180°: they now are opposing each other, and now the two sounds cancel.

Figure 13.4 should help to visualize the situation. At the time 0 there is a maximum for both oscillations #1 and #2. As time goes by the maxima of oscillation #1 occur before those of #2. Eventually, the maxima of oscillation #2 arrive so late that they coincide with the minima of oscillation #1 (position A on the time line of Figure 13.4). They are now 180° out of phase. As we carry on the drawing, the maxima of oscillation #2 lag further and further behind, until one full cycle is missed; and a maximum in #2 happens at the same time as the "next" maximum of #1 (position B). The two oscillations are 360° out of phase, but that is the same as being in phase again!

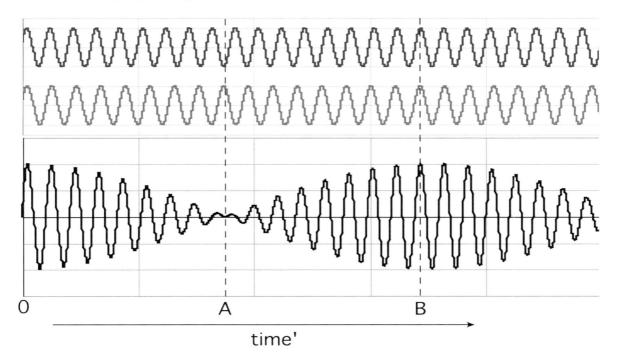

Figure 13.4. Oscillations #1 and #2, with slightly different frequencies. During the time the first oscillation (blue) makes 18 full cycles, the second one (magenta) makes 17. The sum of the two oscillations (bottom) shows "beats".

The term "beats" refers to the acoustic effect of alternating loud and soft sound coming from the alternating in-step and out-of-step oscillations. The "*beat frequency*" is the repetition rate of loud and soft alternations, i.e. how many times a second the two sounds cancel.

In general, if one adds two oscillations with frequencies f_1 and f_2 and corresponding amplitudes A_1 and A_2, these <u>add</u> most strongly $(f_1 - f_2)$ times per second. At these times the resulting amplitude will be $A_1 + A_2$. Also they subtract from each other $(f_1 - f_2)$ times per second, and at these times the resultant amplitude will be $A_1 - A_2$ (or $A_2 - A_1$, whichever is positive).

The beat frequency is equal to the difference in the frequencies of the two sources:

Beat Frequency:
$$\boxed{f_B = (f_1 - f_2)}$$ **(13.2)**

Beats between piano strings: if you open up a piano you will notice that the same hammer strikes two (or on a grand piano even three) strings. When a piano is badly out of tune, there will be rapid beats between the two strings producing the rough, unpleasant sound normally associated with a honky-tonk barroom piano. On the other hand, in tuning a concert grand piano, the tuner may well choose to allow a very small mistuning so that the sound that lingers has a soft, long modulation in amplitude.

13.2.3. HARMONY

Certain tones played together sound pleasant to us, harmonious, while others sound rough, dissonant. This is not to say that dissonant sounds are not used in music - it is often the contrast between harmony and dissonance that makes a composition interesting.

If you hit two keys on the piano at the same time, certain combinations sound "in tune" or harmonious. These combinations are given special names, depending on how many keys apart they are, as we will see later when we talk about musical scales. For now we just want to mention a few of them and explain what is special about harmonious intervals.

The frequency relationship between two tones is called a *musical interval*. An interval is characterized by the *frequency ratio* of the two tones. Some harmonious intervals ("consonant" intervals as opposed to "dissonant" intervals) are listed below:

name of interval	frequency ratio	Example
unison	1:1	300 Hz + 300 Hz
octave	1:2	300 Hz + 600 Hz
fifth	2:3	200 Hz + 300 Hz
fourth	3:4	150 Hz + 200 Hz

You note that the *frequency ratios* in all these cases are what you might call *simple-number ratios*, like 2:3 rather than for instance 23:31. The superposition of two tones of ratios like those shown in the table above has a special property: the sum is periodic and repeats itself after a short time. You can look at it this way: for a frequency ratio e.g. 2:3, in the same time interval in which the 200 Hz tone does two full oscillations, the 300 Hz tone does three full oscillations, as illustrated in Figure 13.5. Notice that the superposition of the two tones is a non-sinusoidal but periodic with a period of 10 ms. Why 10 ms? The periods of the two tones (200 Hz and 300 Hz) are 5 ms and 3.33 ms, and we must wait long enough for <u>both</u> to have completed an integer number of oscillations. After 10 ms, the two oscillations have completed 2 and 3 full oscillations respectively. The repetition period is thus 10 ms = 0.01 s and the repetition frequency is f = 1/T = 100 Hz.

In general, the frequency with which the superposition tone repeats itself is the *largest common divisor* of the two individual frequencies. The largest common divisor of 200 and 300 is 100. A number like 50 would be a common divisor also, but 100 is the <u>largest</u> one.

Figure 13.5.

200 Hz Signal

300 Hz Signal – Note that both the 200 and 300 Hz signals are zero at 0 ms, 10 ms and 20 ms.

The sum (superposition) of the 200 and 300 Hz tones. The period of the sum is 10 ms.

13.3 FOR THE FUN OF IT: LISSAJOUS FIGURES

Lissajous had the idea to graph two oscillations in a different way: show the displacement that corresponds to one oscillation along the x-axis while the other displacement is shown along y. You can see a simple demonstration of this in the Exploratorium in San Francisco, where a pendulum like that in Figure 13.6 hangs from the ceiling. The pendulum has a different period for x and y motion. The curve traced out by the paint bucket (or by a felt pen) is a Lissajous figure.

Figure 13.6. A Lissajous pendulum.
The pendulum swings left-to-right faster than back-and-forth because it is shorter in one direction than the other. If you make it oscillate left-to-right and back-and forth at the same time, the blue paint dripping down from the paint bucket traces out a curve called a Lissajous figure.

What do we need Lissajous figures for? We don't really need them, but they are fun to physicists, and perhaps also to some graphic artists (see figure next to chapter title). Furthermore, as we shall see shortly, they can be used in the lab to tune two tones to be exactly in a harmonious ratio. In this case we would use an oscilloscope and connect one signal to the x-input, the other to the y-input so that the dot on the face of the scope gets deflected accordingly.

Let us look at some examples:
(a) Same frequency: this would correspond to the oscillations of a normal pendulum (without the V in the string of Figure 13.6) which oscillates along x or along y or any other direction with the same frequency.
The graph you get depends on the phase shift between the x- and y- oscillations. Think about it: if the x- and y-oscillations are in step, the pendulum traces a straight line. If both amplitudes are the same, the straight line is at 45°, as shown in the first graph below. Now if you imagine the two oscillations being out of phase by 90° (a quarter of an oscillation) as in Figure 13.2, you see immediately that when one displacement is at the maximum the other is zero: the pendulum moves in a circle. For phase shifts intermediate between 0° and 90° the path of the pendulum is an ellipse. For 180° phase shift it is again a straight line. Note that all of these are curves that come back upon themselves after one oscillation, because the period in x and y displacements is the same.

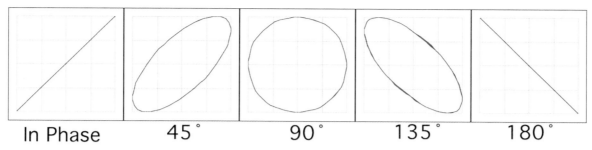

In Phase   45°   90°   135°   180°

Figure 13.7. Lissajous figures for the frequency ratio 1:1.

b) Frequency ratio 2:1. If one oscillation is twice as fast as the other (octave interval in music) the pictures again depend on the relative timing of the two oscillations, i.e. if they cross(zero at the <u>same</u> time as was shown in Figure 13.1, or are shifted with respect to one another. In contrast to Figure 13.1 here we assume again that the amplitudes in x and y are the same so that the figure fits into a square. Try to figure out which of the figures below corresponds to the relative timing of the oscillations in Figure 13.1.

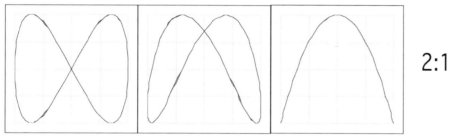

2:1

Figure 13.8. Lissajous figures for 2:1 frequency ratio

Whether you think this is fun is a matter of taste. But it would be good to realize that *Lissajous figures are very useful for exact tuning to a simple frequency ratio.* If the frequency ratio differs ever so slightly from the exact ratio, the figure does not repeat itself again and again, rather the figure changes slowly from one of the patterns to the next.

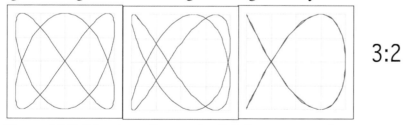

3:2

Figure 13.9. Lissajous figures for 3:2 frequency ratio.

(c) What is the frequency ratio? Try to figure out on your own what the frequency ratio is for the set of pictures shown in Figure 13.10. You can get a hint if you ask yourself why the curve for the 1:2 ratio touches the top and bottom of the graph twice while it touches the left and right side only once.

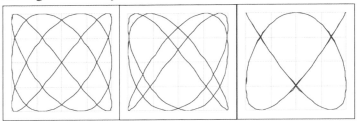

What is the Frequency ratio?

Figure 13.10.

(d) Non-simple ratios: for a frequency ratio like 1:1.571 almost the entire square will be filled with ink before the line closes. You may remember from math that some numbers cannot be expressed as a ratio of two whole numbers. For such "irrational" relationships between the two frequencies, the curve never closes – the entire page will eventually get filled with ink.

Chapter 14

SOUND WAVES

14.1 PROPAGATION OF A PULSE

If you drop a stone in a still pond or an undisturbed swimming pool, a disturbance is created on the surface of the water. A series of sketches of what happens is shown in Figure 14.1.

One sees that the disturbance, or bump on the water surface, travels across the pond.

Speed of propagation:
One can measure the speed with which the bump travels across the water. To find this speed divide the distance traveled (D in Figure 14.1) by the time (t_2 - t_1) it took to travel that far:

$$v = \frac{D}{(t_2 - t_1)}$$

Figure 14.1. A rock is dropped in a pond

If you put some leaves or paper boats on the water you see that they bob up and down as the disturbance moves under them. They *do not* move along the surface with the bump, rather, the water moves only up and down. Energy of motion is being transmitted across the pond by the water without any water actually moving from one side of the pond to the other. The same with sound propagating in air: the air molecules push on one another and transmit the sound from place to place without any motion of air from the instrument to the listener.

The same phenomenon of a traveling disturbance can be seen on a stretched slinky or on a taught, long rope or string: shake one end sharply and you can see a bump traveling along it. Exactly the same happens on the taught string of a piano or a string instrument. To understand vibrations of strings we need understand how to figure out the speed of propagation.

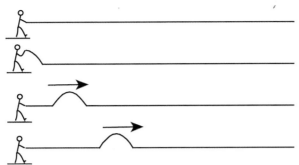

Figure 14.2. Transverse pulse on a string or a rope.

The height of the bump obviously depends on how strongly one has shaken the end of the rope. The velocity of propagation of the bump does <u>not</u> depend on <u>how</u> the end was shaken: it depends only on the *tension* on the string and on *mass per unit length* of the string

Speed of propagation

$$v = \sqrt{\frac{F}{M/\ell}} = \sqrt{\frac{F}{\rho}} \qquad \textbf{(14.1)}$$

Here F is the tension on the string (in Newtons), i.e. the force applied at the ends of the string to keep it taught, M is the mass of the string (in kg) and ℓ is the length of the string (in m). What matters to the speed of motion is the quantity M/ℓ, i.e. the ***mass of 1 meter of string***. Whether you cut a long or a short piece of string to figure out M/ℓ does not matter, because M and ℓ increase in direct proportion to one another. So sometimes it may be better to call M/ℓ by a new name, like Greek rho: $\rho = M/\ell$ = mass per meter length of the rope. Usually, the symbol T is used for the tension, but to avoid confusion with the period T we simply call it F.

Does the formula (14.1) for the speed of propagation make sense? To understand the formula, we look at the propagation of a pulse in detail. For our "rope" we will use a large number of balls strung like beads onto a strong thread. Then we stretch the thread with a tension, F, hit one ball with a hammer and watch the pulse propagate as in Figure 14.3.

Intuition will tell you that when the first ball is hit downward by the hammer, the second ball also moves downward but a little <u>later</u> than the first ball. Why? - Because after the first ball has moved a small distance, the taught string starts pulling on the second, but it takes time for this to pick up speed.

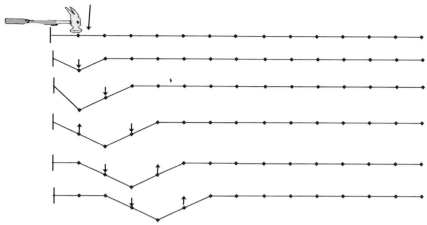

Figure 14.3. The dots represent balls on a taught, stretched string. The small arrows indicate the direction of motion of the balls.

For the same reason the third ball will move only after the second ball moves and with the same delay. The accumulation of delays, combined with the separation of balls yield a certain speed for the pulse propagation. This speed is higher if the <u>delay</u> is smaller and vice versa. Since heavy balls move "sluggishly" for a given force, the delay is longer for heavy balls than for light balls. Similarly, if the balls are closer to each other, there is more mass to be set in motion and the propagation is slower. So the **greater the mass per meter length, M/ℓ, the slower the pulse** travels. It thus makes sense that M/ℓ in the formula is in the denominator. If we increase the tension, F, the force on the second ball due to the first is larger, it will pick up speed (accelerate) faster and the delay is shorter. So **greater tension F means faster pulse speed v**. Again the formula works out. The square root sign is harder to understand intuitively, but without the square root the units do not work out, just as we saw in Chapter 11 for simple harmonic motion.

To summarize:

> *The greater the tension the higher the speed of propagation*
> *The heavier the rope or the string the slower the speed of propagation.*

A word of caution: The formula (14.1) is true only if the string is very limber, if it has no stiffness. Any stiffness of the string adds an additional force (besides the tension) which is not taken into account in the formula.

14.2 LONGITUDINAL AND TRANSVERSE WAVES

If one end of the rope (or of a long slinky) is moved smoothly up and down at a constant rate, a *wave* travels along the string. You can do this by stretching a long slinky on the floor and moving one end smoothly back and forth again and again. Figure 14.4 shows the shape of the slinky at different instances in time. A wave is different from a simple oscillation: a wave extends in space with many points along the wave oscillating at the same rate same rate. The waves we discuss here are mechanical waves, quite different from light waves which, as we learned, are electromagnetic waves.

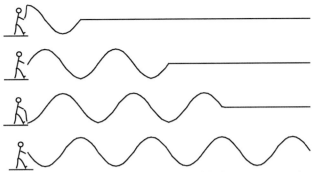

Figure 14.4. Transverse sinusoidal wave on a long string

Note that in the drawing of Figure 14.4 the wave travels from left to right, but each point of the string moves up and down, i.e. the motion of the string is at 90° from the direction of propagation. This type of wave is called a *transverse wave*.

Now suppose you move the end of a slinky <u>along</u> the slinky axis, rather than transverse to it, i.e. you push and pull on the slinky again and again to compress and to stretch it periodically. Now the direction of motion of each part of the slinky is the same as the propagation direction (Figure 14.5): This is called a **longitudinal wave**.

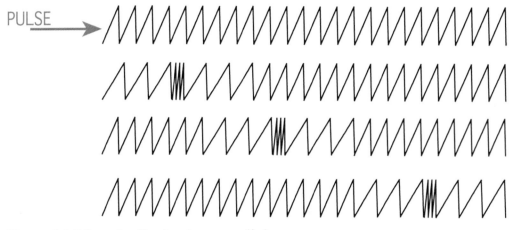

Figure 14.5 Longitudinal pulse on a slinky.

14.3 SOUND WAVES IN AIR ARE LONGITUDINAL WAVES

Air is squishy- air can be compressed, as you know if you ever used a simple bicycle pump. Sound waves in air are longitudinal waves - a big loudspeaker membrane that moves rapidly back and forth making alternate compressions and decompressions can create sound waves. When compressions are passed to the neighboring air particles, they are set in motion because the compressed air near the speaker expands and pushes on the nearby air particles much like we visualized for longitudinal waves on a slinky.

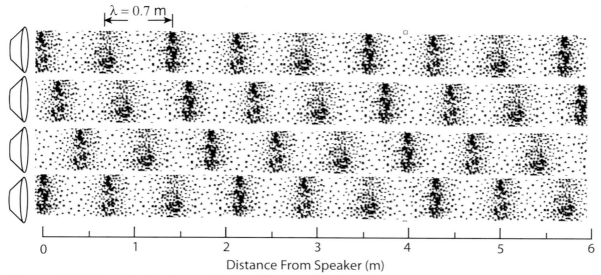

Distance From Speaker (m)

Figure 14.6. Longitudinal sound wave from a loudspeaker. The wave moves left-to-right and is shown at different instances in time. Greater density of dots indicates greater air pressure. The distance from one wave crest (maximum density) to the next is called the wavelength λ. The bottom drawing is lagging one period T behind the top drawing.

In Figure 14.6, the dots symbolize the air molecules. At a wave crest the air is denser, at a wave trough the air is more dilute. You should imagine the crests and troughs to be moving to the right, assuming a loud speaker continues to produce compressions at the left. The picture should not be taken too literally - obviously, there are many more molecules than we can show, but also the change in air density from one place to the next is tremendously exaggerated. Even for the very loudest sounds you can barely stand, the change in air density between the wave crests and troughs is only 1/10,000 of the normal air pressure! So if normal air pressure (no sound) had 10,000 little dots on one area of Figure 14.6, at the peaks there would be 10,001 dots and at the troughs 9,999 dots - you certainly would not be able to notice. Your eardrum is just terribly sensitive: an ant walking on your eardrum makes a deafening sound!

The distance from one wave crest to the next is called the *wavelength* λ (Greek letter lambda). The wavelength is marked in the Figure 14.6. . We encountered the concept already in Chapter 1, but for audible sound the wavelengths turn out to have everyday dimensions from inches to yards, while for visible light the wavelengths are less than a nm.

Making a drawing like Figure 14.6, , full of little dots has two shortcomings. It is tedious to draw, and it is not *quantitative* (i.e. it is difficult to read or measure how compressed or rarefied is the air at a particular point in space and time).

A more convenient way of depicting the sound wave is to draw a series of graphs showing the pressure at various points for different instants in time (Figure 14.7). One sees that the pressure variations are sinusoidal and move to the right. The drawing in Figure 14.7 corresponds to the propagation of a pure tone of frequency 500 Hz in air.

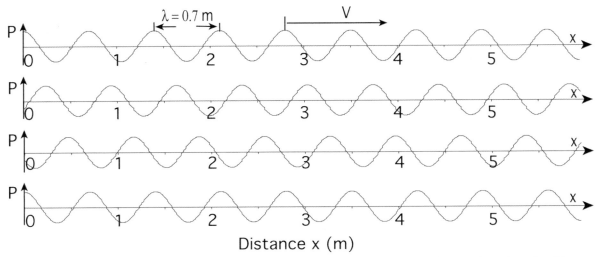

Figure 14.7 Graphs corresponding to the pressure distribution of the drawing in Figure 14.6. With time the wave crests move more and more to the right.

Why can't you produce transverse waves in air? If you were to hold a stiff piece of cardboard in front of you and move it left and right, it would simply slide through the air without making a wave because the air particles are not attached to one another as they are in a slinky. But if you were to move the same cardboard fast enough toward and away from you, the air would be compressed and rarefied, and a longitudinal wave would propagate.

14.4 SPEED OF SOUND IN AIR

To understand wind instruments of all kinds, we need to know the speed with which the pressure waves in the instruments propagate. The speed of propagation depends on the properties of the gas. If the molecules of the gas are relatively heavy, they move more slowly than for lighter molecules. Consequently, it takes longer for the compression to reach the neighboring molecules and thus sound propagates slowly. For example in helium, whose atoms are much lighter than air molecules, the speed of sound is almost 3-times greater than in air

With increasing temperature the gas molecules move more rapidly and the speed of sound increases accordingly. The kinetic energy of the gas molecules, $(1/2)$ mv^2, is proportional to the absolute temperature: absolute zero temperature is at -273.2°C (-460°F), and at this temperature all molecular motion stops. To figure out the change in speed of sound for instance between 0°C (freezing point of water) and 20°C (room temperature, 68°F), you would add 273.2°K to get the absolute temperatures: 273.2°K and 293.2°K. Now the ratio between these two numbers is 0.932. This is the ratio of molecular kinetic energy, i.e. the ratio of the square of velocity at the two temperatures. To find the ratio of velocities, you need to take the square root of 0.932, which is 0.965. So if the speed of sound at 20°C is 344 m/s, at 0°C it is 0.965 × 344 m/s = 332 m/s.

To save you the trouble of doing this calculation for different temperatures, here it is, giving you the two important numbers: the speed of sound in air at room temperature, and the change in speed of sound for every degree change in temperature:

> *speed of sound in air at room temperature (20°C) v = 344 m/s*
> *increase in v with temperature: 0.17% per °C = 0.6 m/s per °C.*

In U.S. customary units (speed in feet/s and temperature in °F) the corresponding numbers are:

> *speed of sound in air at room temperature (68°F) v = 1130 ft/s*
> *increase in v with temperature: 1.1 ft/s per °F.*

The change of speed of sound with air temperature is important for brass and woodwind musicians. The pitch (frequency) of the instruments depends on the speed of sound - and therefore depends on the temperature of the air inside the instrument. As we will see later when we talk about oscillations of wind instruments, the frequency changes in direct proportion with the speed of sound. A wind instrument gets warmer after it is played a few minutes - thus, when a flutist warms up her instrument for instance by 10°C, (18°F) the speed of sound increases by $10 \times 0.17\% = 1.7\%$. This raises the pitch very noticeably.

14.5 WAVELENGTH AND FREQUENCY

The wavelength is the distance from one wave crest to the next, as shown in Figure 14.7. But you could just as well have used the distance from one wave trough to the next, or the distance from any one point of the wave to the next point which has the same phase. A wavelength simply includes one complete oscillation.

The rate at which the wave crests arrive at the listener depends on two things: on ***the speed v*** with which the wave travels toward the listener and the ***wavelength*** $\square$. You will easily be able to figure out the relationship between frequency and wavelength. If the sound source makes f oscillations in one second, and the wave travels v meters away from the source in one second, there will be f oscillations distributed over v meters, so the wavelength must be:

$$\text{wavelength: } \lambda = v/f. \qquad \textbf{(14.2)}$$

Note that the units in equation (14.2) work out: v has units of *m/s*, and f has units of *1/s*, so *v/f* has units of *m*. The equation says that if the oscillation frequency is high the wavelength is small - well yes, if you move your big toe in the bath tub back and forth rapidly, the wave crests are close together, because you make more ripples than if you moved the toe slowly. Instead of memorizing the equation above, you can always figure it out quickly.

Example: find the wavelength in air of a very low-pitched sound (100 Hz) and a high-pitched sound (10,000 Hz = 10 kHz).

Answer: for 100 Hz: $\lambda = \dfrac{v}{f} = \dfrac{344 m/s}{100/s} = 3.44 m$

For 10,000 Hz: $\lambda = \dfrac{v}{f} = \dfrac{344 m/s}{10,000/s} = 0.034 m$

In U.S. customary units, for bass tones, the wavelength is about 10 feet, while for a high-pitched tone the wavelength is just over an inch.

Relevance to size of instruments or loud speakers

The above example shows that bass tones have a long wavelength. To produce a strong bass tone, the instrument or loudspeaker needs to be large. A vibrating violin string by itself will hardly make any sound, because as it vibrates back and forth, when air is pushed away as the strings move one way, the air simply goes around to fill the void on the back of the string. You best see this if you hold a pencil vertically in the water and move it back and forth. For a slow motion the water has time to flow around the pencil and makes hardly any waves, only for a very rapid motion, when the waves have a very short wavelength does a pencil make noticeable waves. To make waves with a **long wavelength** you need a **large object**, like a canoe paddle! The conclusion: the size of the sound source (the "radiator") must not be much smaller than the wavelength of sound that is to be radiated. There is an English saying: you can't fan a fire with a knitting needle. Thus a bass fiddle or a bass speaker needs to be fairly large. A violin radiates higher frequencies and thus can be smaller. Bass speakers increase the efficiency of sound radiation at low frequency by suitable enclosures that prevent air that is pushed away at the front of the speaker from going around to the back of the speaker filling the void in the back, rather than projecting the sound out of the front.

14.6 SOUND PROPAGATION

1. Reflections: sound waves disperse through rooms and to adjacent rooms by multiple reflections. Even if there are obstructions between you and your radio speaker, you hear the sound reflected from the walls or the ceiling (Figure 14.8).

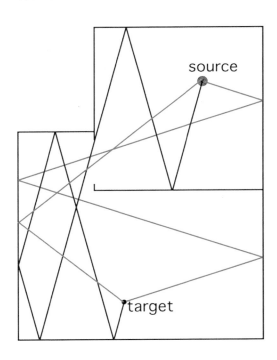

Figure 14.8. Three possible paths from a source of sound to a target (a person) are shown. The three paths shown are not all possible paths. Shown are paths with 1, 3 and 6 reflections respectively.

2. Diffraction and shadowing

No doubt you have noticed that another person standing in front of a table lamp makes a shadow on the book you are reading, but does not obstruct the sound waves from your table radio. Why are light waves more easily obstructed than sound waves?
The answer is that light waves have very short wavelength while sound has wavelengths a few inches to a few feet.

Shadowing of sound or light waves depends on the size of the object compared to the wavelength. If the wavelength is short compared to the obstacle, there is a shadow behind the obstruction (Figure 14.9). This is the usual case for light waves, because the wavelength of light is so short that normal objects are always much larger than the wavelength of light. But sound waves of lower frequencies have rather large wavelengths, which bend around the obstacle (Figure 14.9).

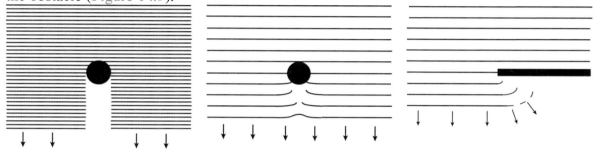

Figure 14.9 Wave propagation near an obstacle. If the wavelength is much shorter than the size of the obstacle, a shadow is created (left figure). The larger the wavelength compared to the size of the obstacle, the more the waves bend around the obstruction (diffraction of waves, pictures on the right).

This would suggest that if you sit behind a big stone pillar for a concert in a gothic church the high frequencies would be completely shadowed. If the pillar has a diameter of 1 m, wavelengths much shorter than 1 m would be shadowed. But a 1m wavelength corresponds to 344 Hz, i.e. much above 344 Hz would not be heard. So almost half of the tones on a piano keyboard would not be heard. In reality, even if you sat behind the pillar, where you get very cheap tickets, you still hear the higher frequencies as reflections from the many other surfaces where sound is reflected (Figure 14.8).

14.7 INTERFERENCE OF SOUND WAVES

The term "interference" of sound waves is commonly used to describe the superposition of sound waves of the same frequency from two sources at different distances from the listener. The term is misleading because the oscillations caused by different sound sources do not really hinder one another, as we see with water waves in the figure next to the title of this chapter. The two waves simply add as we have learned earlier when we discussed superposition.

As an interesting illustration we can imagine stereo loudspeakers that are placed some distance apart, as shown in Figure 14.10. Suppose that your amplifier is switched to "mono", which means that both speakers are connected to exactly the same signal and

oscillate in phase (in step). What will be the result of the superposition of the two sound sources? At first you might say that they will simply add to one another, since they are in step. But no - we must take into account that we may be closer to one speaker than the other, so that the wave from one speaker arrives a little later than the one from the other one. You might say that sound travels pretty fast and so the delay would not amount to much, but on the other hand sound oscillations are pretty fast, so the delay should always be taken into account.

What will happen? To think of a specific example, assume that both speakers emit the same pure tone of 1100 Hz, for which the wavelength is about 0.3 m. If you are the same distance from both speakers, the waves arrive in step (in phase) because the delay from the speaker to your ear is the same. But if the distance differs, then one wave will be delayed with respect to other. If the delay is just half an oscillation, the two waves tend to cancel - they make "destructive interference", as shown in the lower picture of Figure 14.10. This is the case when the difference in distances from the two sources is exactly one-half wavelength, in our example 0.15 m. But come to think of it: what if you move your ear more to the side, so that the distances from the two speakers is bigger? Well, when the delay is one whole wavelength the two waves will reinforce again just as in the middle - we have constructive interference. Thus you can well imagine that as you move off to the side, there will be alternate places of loud and soft sound. Loud every time the difference in distance to the speakers is 1 or 2 or 3 whole wavelengths, soft whenever the difference is 1/2, 3/2, 5/2 etc wavelengths.

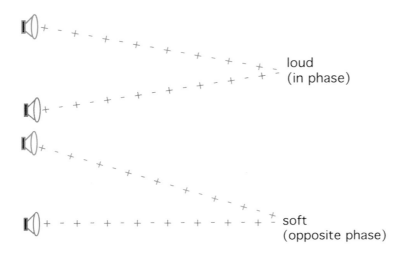

loud
(in phase)

soft
(opposite phase)

Figure 14.10 Interference from two loudspeakers. The + and – signs indicate wave crests and wave troughs. When + meets + and – meets -, the sounds reinforce (as shown in Figure 1.2). In the figure at the bottom here, the listener moved to the side. Now the distance to the two speakers is no longer the same. If the difference in distance equals 1/2 or 3/2 or 5/2 wavelengths, + meets – and the sounds tend to cancel one another.

In a room the effects are not so pronounced because there is also sound coming from wall reflections. But still there is a practical application to test whether your stereo speakers are correctly connected. For "mono" the speakers are supposed to oscillate together in phase - but if you have one pair of wires crossed they are 180° out of phase. To do a test, put one of your ears near the speakers, the same distance from both speakers and cover your other ear to reduce hearing room reflection. If you have a friend reverse the pair of wires on one speaker while you listen, the sound will be loud for the correct connection, and soft for the incorrect one, since for the incorrect connection the speakers are opposing one another.

Chapter 15

SOUND PERCEPTION:
PITCH, LOUDNESS, TIMBRE

Now that we have learned about the basic properties of sound, let us begin studying what effect sound has on our ears.

For the present purposes, it is more convenient to use electronic circuits for sound production rather than a musical instrument, because we can more easily control and measure frequency, amplitude and wave shape. Let us connect a loud speaker (or headphones) to the signal generator (an electronic "oscillator"), and display the wave shape on an oscilloscope.

It is easy to discover relationships between physical stimulus (what you observe on the oscilloscope) and the perception of the sound. Amplitude of the sound waves corresponds to the perceived *loudness*, while the frequency is perceived as *pitch*.

If one changes the shape of the wave from sinusoidal to triangular or square wave, the timbre is changed, but the pitch is not. *Timbre* (pronounce it as in French, or as in the first two syllables of the word tambourine) is the "quality" of the sound. If you listen to a flute or a violin, both playing the same tone (same pitch) with the same loudness, you can immediately tell that they sound different. The aspect that distinguishes them is called the timbre, or, by analogy with light, the physical "tone color". The timbre is associated with the *wave shape* of the sound pressure variations, and this differs between violin, flute, and any other string or wind instruments playing the same tone. We will later see that the wave shape of the tone is determined by its harmonic content. The way the tone rises in time (the attack) and the way the oscillations diminish when the tone ends is also of much importance. These relationships can be shown schematically as:

stimulus		perception
amplitude		loudness
frequency		pitch
wave shape		timbre

The black lines suggest that loudness depends not only on amplitude but also on frequency, and for the same frequency very loud and very soft tones are perceived by some people as having slightly different pitch. More detailed experiments on human hearing suggest that to a small extent all three aspects of the stimulus affect all three aspects of perception, but red and solid black lines in the diagram above represent the most important aspects of tone perception.

15.1 LOUDNESS AND AMPLITUDE

No doubt, when one increases the amplitude of the pressure variations by turning up the volume control on the radio the perceived loudness increases.

But does the sound get twice as loud when you make the amplitude twice as large? This question reveals a fundamental problem of perception studies: how would you know exactly what one means with "twice as loud" - and would tests on different people who are asked to judge "twice as loud" in an experiment give similar results?

A much more objective test on loudness perception is done playing two tones of the same frequency (e.g. 1000 Hz) but slightly different amplitude one after the other. The difference in amplitude is then made smaller and smaller until the listener can no longer tell the difference. This is done by alternating between the softer (S) and the louder (L) tone randomly (L L S L S S S L S etc.) and the listener indicates which tones sound louder or softer. In this way one finds the JND (just noticeable difference) in loudness perception.

The results are surprising in two respects:
(a) We are rather insensitive to small differences in amplitude: increasing the amplitude 1.5-fold is only barely noticeable.
(b) A given increment in loudness corresponds to a certain ratio of sound pressure amplitude. For instance, a change in amplitude from 10 to 15 units, to us sounds the same as the change from 20 to 30 units, or from 100 to 150 units. Note that from 10 to 15, the amplitude changed by 5 unit, from 100 to 150 the amplitude changed by 50 units, yet subjectively the loudness change is the same in both cases. The same is observed for louder tones: **equal increments in loudness** corresponds to **multiplying the amplitude by the same number.** Many audio systems let you increase the volume in steps. These are equal steps in loudness, each time multiplying the amplification by the same factor.

Another surprising aspect of our loudness perception is the vast range of amplitudes that we can perceive, from the softest tone to the loudest tone that we can tolerate. Sound pressure amplitude is measured in N/m^2 - it tells how much force (in Newtons) is pushing against the eardrum per unit area (pressure = force/area). For a frequency of 1000 Hz, for the average person of normal hearing the threshold of hearing is a sound pressure amplitude of 0.00003 N/m^2 = 3×10^{-5} N/m^2. Compare this pressure to that exerted by a single sheet of paper on the table: the threshold of hearing is 20,000-times less pressure! But at the other extreme, the loudest sound you can still stand - although painful - exerts a pressure 1,000,000-times greater than the threshold of hearing, i.e. a pressure of 30 N/m^2 (50 sheets of paper).

Sound Intensity means how much sound energy hits one square meter of area that absorbs the sound. Energy is measured in Watt (abbreviated W, such as in 60 W light bulb). *Sound intensity (in W/m^2) is proportional to the square of the sound pressure amplitude.* Without going into details, a similar relationship holds between the amplitude and energy of a pendulum or a mass oscillating on a spring: doubling the amplitude quadruples the energy, because for twice the amplitude the spring gets stretched twice as far but in addition is twice as hard to pull: $2 \times 2 = 4$, i.e. the square.

How large is the sound intensity for the loudest sound you can tolerate compared to the threshold of hearing? The sound pressure amplitude is 1,000,000-times larger, so the **sound intensity** is the square of that, or **1,000,000,000,000**-times larger, or 10^{12}!!

The table below shows some of the important aspects of what we just talked about. The table lists different typical intensities of sound sources, from a jet engine to the softest sound you can barely hear in a sound proof enclosure (threshold of hearing). In each case we list the sound pressure amplitude, the ratio of the sound pressure to the threshold sound pressure (p_o) for a 1000 Hz tone, and the corresponding intensity level I in W/m^2. The next column, labeled I/I_o, gives the sound intensity relative to threshold intensity I_o.

For the loudest sound, the intensity is, as mentioned, 10^{12}-fold larger than the threshold intensity.

DESCRIPTION & MUSICAL NOTATION	PRESSURE AMPLITUDE p in N/m^2	p/p$_o$ press relative to threshold	INTENSITY I in W/m^2	I/I$_o$	SOUND LEVEL in dB
Jet engine at 20'	30	1,000,000	2		
Loud as Hell	3×10^1	10^6		10^{12}	120
Rock Band	3	1,000,000	0.02		
Fortissimo (fff)	3×10^0	10^5	2×10^{-2}	10^{10}	100
Very Heavy Traffic	0.3	10,000	0.0002		
Loud f (forte)	3×10^{-1}	10^4	2×10^{-4}	10^8	80
Quiet Conversation	0.03	1,000	0.000002		
Soft (p - piano)	3×10^{-2}	10^3	2×10^{-6}	10^6	60
Quiet room	0.003	100			
very soft (ppp)	3×10^{-3}	10^2	2×10^{-8}	10^4	40
Whisper at 4'	0.0003	10			
Hard to hear	3×10^{-4}	10^{-1}	2×10^{-10}	10^2	20
Threshold of hearing	0.00003	1			
at 1000 Hz	3×10^{-5}		2×10^{-12}	1	0

Remember that equal increments in subjective loudness correspond to equal **factors** of sound amplitude or sound intensity. For each step in the table, the sound pressure increases 10-fold, and the intensity, being the square of the sound pressure, increases 100-fold.

It seems that a better measure of loudness is, therefore, the number of zeroes in the intensity level relative to threshold. The intensity of the loudest sound (relative to threshold) has 12 zeroes, the softest (threshold) has none. This indeed is how the physical sound intensity level is specified, except that the number of zeroes is multiplied by 10. **This is, by definition, the sound level in decibels (abbreviated dB):**

Sound intensity level in dB

$$dB = 10 \log \frac{I}{I_0}.$$

(15.1)

$\text{Log}_{10}(I/I_0)$ is a fancy mathematical way of indicating that we take the number of zeroes in the I/I_0 ratio. Calculated this way, then, the loudness in the last column of the table shows that the loudest tone has 120 dB, the threshold is 0 dB, and in between equal increments in the number of dB correspond to equal increments in subjective loudness.

One way you use this equation is to put the number for I/I_0 in your calculator and then press the "LOG" button. Try it – if you put in 100 you should get the answer 2, because there are two zeroes.

We will be using logarithmic scales again later when we talk about musical scales. It may be useful to show how numbers on a logarithmic axis are distributed:

In contrast to the normal number line, where equal increments in the number are represented as equal distances, here **equal factors** are represented by **equal distances**. Note for example that a 10-fold increase (from 1 to 10, or from 5 to 50, or from 10 to 100) shows up as the same length. Perceived loudness increases the same amount for every 10–fold increase in sound intensity level – thus you can think of perceived loudness represented by the distance along a log number line, while the numbers along the line correspond to the intensity level.

We mentioned earlier that we can not very sensitive to small changes in sound intensity level. Under good laboratory conditions, a 1.5 dB change is just noticeable, but in real life a 3 dB change is needed before you call a tone clearly louder. To calculate I/I_0 for 3 dB, you put 3 into the calculator, divide by 10 and press the inverse function of log, called 10^x. The result you get is 2.0. This tells you that doubling the sound intensity level is needed to bring an appreciable increase in subjective loudness.

If you have a good hi-fi amplifier the volume control may be labeled "loudness" and the numbers given on the dial or the digital display are the dB numbers. Sound intensity levels in dB are always with respect to some arbitrary reference. In equation (15.1), the reference level is the threshold of hearing, but dB is a useful number to compare <u>any</u> two intensity levels. On a hi-fi amplifier, the reference level is rather arbitrary, often chosen for the very highest amplification to be called 0 dB and to specify lower amplification by negative numbers.

15.2 LOUDNESS AND FREQUENCY

A simple experiment demonstrates that perceived loudness depends not only on amplitude, but on frequency also. If one slowly changes the frequency of a tone with fixed amplitude from 1,000 to 30,000 Hz, the perceived loudness of the sound decreases sharply in the neighborhood of 15,000 Hz. Above 22,000 Hz one hears nothing, although a microphone and oscilloscope would confirm that there is a sound wave present. The air is transmitting pressure waves, but the frequency is too high for our ears to hear - it is "ultrasound". With increasing age, one hears less and less of the high frequencies - already at age 25 the ability to hear very high frequencies declines. At low frequencies, hearing extends to about 20-30 Hz - for large amplitude sounds below 20 Hz you may still perceive some sort of pressure variation but not a tone of a definite pitch.

If we compare sounds with different frequency, how is their perceived loudness related to the intensity, or the amplitude of the sound waves? By switching back and forth between two tones of different frequencies and adjusting the amplitude until they seem about **equally loud**, we can study how loudness perception changes with frequency. Do two tones at different frequencies sound equally loud when they have the same amplitude,, i.e. the same intensity level I in W/m^2?
The answer is: not at all!

The result of such measurements are called the Equal-Loudness curves. To explain the method we show one example in Figure 15.1. One uses as a reference a pure tone of 1000 Hz, and adjusts the sound intensity level to the desired value, in this example 40 dB. The perceived loudness of this tone is called loudness level 40 *phons*. Tones at other frequencies which are judged to sound "equally loud" are also said to have a loudness level of 40 phons. What the measurements show is that the physical sound pressure or sound intensity level is very much different for different frequencies, i.e. the ear has different sensitivity at different frequencies. The graph shows that for instance at 100 Hz, the sound intensity must be 60 dB to sound as loud as a 1000 Hz tone of 40 dB intensity. This difference of 20 dB is very large, if you remember that 20 dB means 100-times greater intensity level (or 10-times larger sound pressure level). The ear is <u>much</u> less sensitive at 100 Hz than at 1000 Hz.

Figure 15.1. Curve of equal loudness: the curve shows the sound intensity level (in dB) that is needed to match the loudness of a 40 dB tone at 1000 Hz. All the tones along the curve have a loudness level of 40 phons.

An interesting result is found when the same experiment is repeated starting with differently loud sounds at 1000 Hz. Figure 15.2 shows an entire set of measurements of perceived loudness as a function of frequency, for different intensities 10 dB apart at 1000 Hz. When the tone is very loud, such as 90 dB, the curve is almost flat, indicating that for very loud sounds the sensitivity of the ear does not depend much on frequency. For loud sounds the ear is equally sensitive to different pitches. But at the other extreme, for very soft sounds, there is a tremendous difference in sensitivity: the sound needs to have more than a 1000-times greater intensity to be barely audible at 100 Hz than at 1000 Hz. The graph shows that hearing of faint sounds is most acute in the range of about 1000 Hz to 6000 Hz.

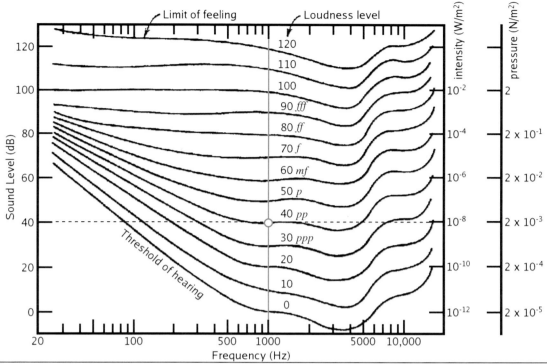

Figure 15.2. Tones at 1000Hz in 10dB increments of sound intensity level are matched in loudness by tones of frequencies between 20Hz and 20,000 Hz. The corresponding loudness levels of musical tones are shown in the usual notation (*f* stands for forte etc). Sound intensity level (dB) on the left axis, intensity (W/m²) and corresponding pressure amplitude (N/m²) on the right. The graph is adapted from H. Fletcher and W.A. Munson, J. Acoustical Society or America **5**, 82 (1933)

You might wonder if any of this is of practical use. One application that will interest you is in music reproduction. Suppose you like to listen to music played at an ear-splitting loudness of 90 dB. Now your neighbors complain and you turn down the volume on the amplifier. While for loud sounds the ear was nearly equally sensitive for different frequencies, after you turn the amplifier down from 90 dB to 50 dB the midrange is still audible, but the lower frequencies have almost completely disappeared, because at low sound intensity levels low frequencies are harder to hear. So you are very unhappy about the sound, unless you turn up the bass quite a lot. Most of the better amplifiers do this automatically, that's why the knob is labeled loudness control and not volume control.

A similar comment applies to the conductor of a band or an orchestra. Since the conductor is close the orchestra, his sound intensity level is much greater than for the audience, so he needs to compensate by "turning up" the volume of the bass and of the high frequencies.

15.3 PITCH DISCRIMINATION

For musicians it is important to know how small a difference in frequency a person can recognize. We learned in Chapter 13 that beats between two tones allow you to detect even the smallest frequency differences when the two tones are played at the same time. Here we ask how well a person can tell when two frequencies are slightly different, one tone played *after* the other.

The answer depends on many factors. Very short tones can be off more than long steady tones – a good thing for the violinist, because if the pitch is a little off-key in a very rapid passage one is less likely to notice than for steady tones. For test tones of one-second duration, experienced musicians can tell when the frequency between two successive ones differs by about 0.3% or 3 Hz for a 1000 Hz tone. So you see that the ear is very good at detecting differences in frequency between two tones, but not good at recognizing differences in sound intensity.

References:
D. A. Hall "Musical Acoustics" Third Edition, Brooks/Cole publishers (2002)
J. G. Backus "The Acoustical Foundations of Music" 2nd Edition, W.W. Norton (1977)T.
D. Rossing "The Science of Sound", 2nd Edition, Addison Wesley (1990)

Chapter 16

THE EAR

The eye and the ear are probably the most complicated sensory instruments in our body - and the most remarkable for their ability to perceive and process a vast range of stimuli. Here we give a brief, simplified review of the way the ear senses sound-pressure variations and converts them into nerve pulses transmitted to the brain. Does the ear work like a microphone, which accurately changes pressure variations into electric signal of the corresponding wave form? How does it manage to work over such a vast range of intensities? What limits the frequency range of the ear?How is it possible for the ear to distinguish tones, which differ in frequency by less than 1%? These are but a few of the questions one might ask when one thinks about the ear as a sound detector. In passing we might mention that the ear not only allows us to hear sound, but it is also the organ that helps us maintain equilibrium.

16.1 THE PARTS OF THE EAR

The components of the ear, shown in Figure 16.1, can be grouped into three parts: *the external ear, the middle ear and the inner ear.*

The external ear consists of the pinna and the ear canal. The pinna is the outer part of the ear, while the auditory canal is the tubular opening leading to the middle ear. Although the pinna, is not as important to us as it is to many other mammals (elephants!), it helps us distinguish the direction from which the sound originates. The pinna alters the intensity of sound from a source behind our head compared to that from a source in front of us. In addition, the pinna collects sound and guides the sound waves more or less toward the ear

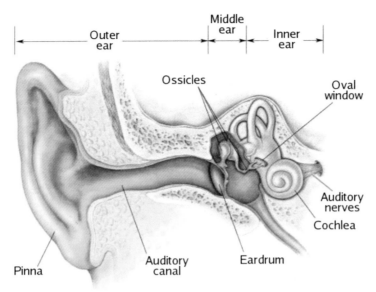

Figure 16.1 The parts of the ear

canal - "more or less" because you remember that in considerations of sound propagation (Chapter 14) one should compare the size of the object to the wavelength of the sound. The pinna is approximately 6 cm in size. The frequency that corresponds to a 6 cm wavelength is about 6,000Hz so that the pinna helps sound collection primarily for the higher frequencies.

The ability to tell from which direction sound originates has two aspects: the difference in sound volume received by the left and the right ear, and the tiny difference in the arrival time of the sound between the two ears,

At the end of the auditory canal begins the middle ear, which consists of the tympanic membrane ("ear drum") and three tiny bones (ossicles) transmitting the motion of the ear drum to the inner ear. These bones are called the hammer, the anvil and the stirrup according to the objects their shapes resemble. These are the smallest bones in the human body.

Together the ossicles act like a lever or a crow bar: the ear drum pushes back and forth at one end (hammer), at the other end (stirrup) the amplitude of motion is less but the force is greater. What is quite remarkable is that our body has the ability to change the effective fulcrum (point of rotation) of the lever, so that for very loud sounds the ear is partially protected from damage. This involves a feedback mechanism in which the brain, when the sound is too loud, makes muscles in the ear change the lever action!

The inner ear contains the cochlea, which converts the mechanical pressure variations into nerve signals. The cochlea is a spiraling, tapered, snail-like, bony structure filled with fluid and has two windows. The "oval window" (Figure 16.1) is attached to stirrup, which wiggles the window back and forth and sets up a pressure wave in the liquid inside the cochlea.

The function of the cochlea is much easier to describe if we imagine the snail-like spiral stretched out, as shown schematically in Figure 16.2. The basilar membrane is a flexible membrane which runs down the middle of the cochlea and divides it into an upper and a lower part, connected by and opening at the apex of the cochlea.

OVAL
WINDOW

ROUND
WINDOW

Figure 16.2. Longitudinal section through the stretched out cochlear duct. The entire cochlea is approximately 35 mm (or 1.5") long. Pressure from the eardrum is transmitted to the stirrup (top left), which presses on the oval window. The pressure wave travels along the top of the basilar membrane through the fluid inside the cochlea and returns along the bottom of the membrane back to the round window on the lower left. In the process, the basilar membrane is deflected, but note that the amplitude of the membrane deformation is much exaggerated in this schematic.

16.2 PLACE THEORY OF PITCH PERCEPTION

Attached to the basilar membrane are about 15,000 nerve cells, which respond to the bending of the membrane. The membrane is much more flexible near the apex than near the oval window. Depending on the frequency of sound, the flexing of the membrane shifts from one region of the cochlea to another, as shown in Figure 16.3. Correspondingly, different nerve cells are stimulated for sounds of different pitches. For example, a 1,600 Hz tone mostly stimulates cells between 10 and 20 mm from the oval window, but almost no cells beyond 22 mm. On other hand, low frequency tones stimulate nerve cells near the flexible, floppy end of the basilar membrane.

The "place theory" of sound perception holds that we recognize the pitch of a tone by the **location** on the basilar membrane of the groups of nerve cells that are affected by the sound. While this plays a role, place theory would be hard pressed to explain why almost anyone can tell apart tones that differ by 1% in frequency, and musically trained people do even better than that. If one imagines shifting the 400 Hz curve by 4 Hz (1% of 400), the shift is so small that it can hardly be the principal way we recognize pitch. Recent studies of signals from single nerve cells provide a better explanation.

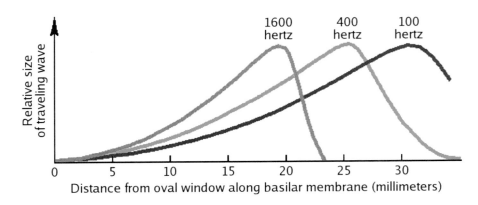

Figure 16.3. For low frequency tones, the basilar membrane flexes mostly near the apex, for high frequency sound nearer to the oval window. Correspondingly, different nerve cells are stimulated for sounds of different pitches.

7.3 WHAT DO THE AUDITORY NERVES TELL THE BRAIN?

The nerve cells in the cochlea send signals to the brain, where the signals are processed into what we call "hearing sound". While the processing in the brain is not well understood, fascinating research on the nerve response of singe auditory nerve cells has been done at the University of Wisconsin Department of Neurophysiology.

In general, nerve cells are known to communicate with the brain by sending electrical signals not in the form of a complex wave shape, but by sending a sequence of individual nerve impulses.

The results of studies on the response if individual auditory nerve cells have found remarkable results, briefly summarized here: .

1. When the auditory nerve is stimulated by a periodic signal, such as a pure tone (sine wave), the nerve pulses occur at various, seemingly irregular intervals. All nerve pulses ("spikes") are nearly of the same height, independent of the loudness of the tone.

2. A histogram of the spacing between adjacent pulses shows an interesting regularity: the pulses are grouped at multiples of the period T of the periodic tone (Figure 16.4).

3. The loudness of the tone changes the **average** rate with which the pulses occur. The rate is related to the sound level in dB of the tone.

4. Nerve pulses are transmitted only during the positive part of the pressure wave, not during the wave troughs.

5. the wave shape is encoded in the detailed shape of the nerve pulse peaks in the histogram of Figure 16.4. For a pure tone, the peaks of the distribution reflect the positive part of the sine wave. For a complex tone, the distribution changes correspondingly.

Figure 16.4. The histogram shows the distribution in spacing between adjacent nerve impulses form a singe auditory nerve cell stimulated by sinusoidal tones of different frequencies. The peaks in the distribution are at multiples of the period of the sound. The histogram for 600 Hz shows that sometimes 10 or more periods pass before the next nerve impulse occurs. The average rate of the nerve impulses (spikes) is related to sound intensity level (dB) The figure is adapted from a paper by Rose et al. in the Journal of Neurophysiology.

When listening to different sounds, the brain receives a sequence of nerve pulses ("spikes") with different time distributions from about 15,000 nerve cells in the cochlea. Since we know that we can distinguish changes in pitch of 1% or less, we conclude that the brain must analyze all these histograms and determine the periodicity of the spike-distribution to high accuracy. Therefore for very soft tones, when the spikes in the average are few and far between, it must take longer to accumulate sufficient information to recognize the pitch accurately. This is indeed the case: the accuracy of pitch perception for very short tones of very low levels is much poorer than for louder and longer tones.

The above brief summary shows that the auditory system encodes loudness, pitch and timbre in a complicated way. We don't fully know how the brain turns this stream of encoded nerve pulses into sensations. You might, however, marvel at your ear and brain functioning, the next time you are aware of hearing sounds.

References

1. Phase-locked Response to Low-Frequency Tones, etc., J. E. Rose, John F. Brugge, David J. Anderson and Joseph E. Hind, Journal of Neurophysiology 30, F69 (1967).

2. Time Structure of Discharges in Single Auditory Nerve Fibers of the Squirrel Monkey in Response to Complex Periodic Sounds, John F. Brugge, David J. Anderson, Joseph E. Hind and Jerzy E. Rose, Journal of Neurophysiology 32, F36 (1969).

3. Some Neural Mechanisms in the Inferior Colliculus of the Cat, Jerzy E. Rose, Nathan B. Gross, C. Daniel Geisler and Joseph E. Hind, Journal of Neurophysiology 29, 288 (1966).

4. Coding of Information Pertaining to Paired Low Frequency Tones in Single Auditory Nerve Fibers of the Squirrel Monkey, Joseph E. Hind, David J. Anderson, John F. Brugge and Jerzy E. Rose, Journal of Neurophysiology 30, 794 (1967).

Chapter 17

VIBRATION OF STRINGS

17.1 SINGLE MODES

The simplest kind of oscillation of a guitar string is a harmonic to and fro motion just like the one shown below for a slinky (Figure 17.1) or a string (Figures17.2). In this case each point of the slinky or the string moves up and down in simple harmonic motion, all points oscillate with the same frequency but with different amplitudes. *When all parts of an oscillating system vibrate in simple harmonic motion the motion is called A MODE of the oscillating system.* The oscillating system could be a string or a drum, and its harmonic oscillation indicates that all its parts vibrate with *the same frequency.* As we will see later, the actual motion when a string is plucked or bowed is more complicated, but we will be able to understand these complicated motions as the simultaneous occurrence of several modes.

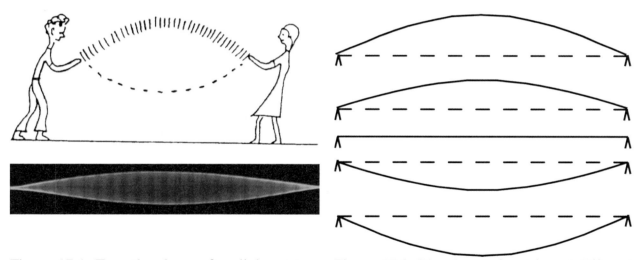

Figure 17.1. Top: the shape of a slinky at two instances in time when it oscillates in the first (fundamental mode). Bottom: photograph of a string in vibration.

Figure 17.2. The shape of a string at different instances in time when the string vibrates in the first (fundamental) mode. The four drawings are an eighth of a period (T/8) apart in time, and represent half a period (T/2).

Only certain special frequencies are allowed. To set up such an oscillation, you would gently wiggle the slinky or the string up and down in sinusoidal motion. The wave you create in this way travels along the slinky and is reflected at the far end where the string is tied down. The wave then travels back toward you and is reflected once more at your end of the string - on an on, the wave goes back and forth. Notice one important point, essential to understand why only certain frequencies are possible on a string: in order to build up a steady oscillation, each subsequent wave crest that you make with your hand should be in step with the wave already present on the string. This means that you should time the wave crests to be spaced exactly equal to the round-trip time of the wave on the string. A whole number of complete waves (1,2,3 ,,,) must fit into one round-trip time. Conclusion: *In the fundamental mode of oscillation of the string, the period T equals the round-trip travel time of the wave on the string.* If the length of the string is called L, and the speed of propagation of the wave on the string is v, then the period of oscillation T (round-trip travel time) is $T = 2L/v$. Since the frequency is the reciprocal of the period T, or $f = 1/T$, then:

First (fundamental) mode frequency: $\qquad f_1 = v/2L \qquad$ **(17.1)**

The formula says that *shorter string corresponds to higher frequency* (string length is in inverse proportion with string oscillation frequency). When the musician shortens the string by placing a finger on the fingerboard (e.g. frets on a guitar's fingerboard), the pitch increases. Note that the speed of propagation v on the string in the formula stays the same as long as the musician does not change the tension (pull on the string) with the tuning pegs.

You can do this yourself: find the round trip travel time of a pulse on a slinky stretched between you and a friend. Then move your hand gently up and down in step with the reflected pulse and observe that you build up a mode that looks like Figure 17.1.

Example. The string of a guitar is 60 cm long and is tuned to standard A = 440 Hz. What is the new frequency, if the vibrating length of the string is shortened to 40 cm by placing a finger on the fingerboard.
Answer: frequencies are in inverse proportion to the string lengths: the ratio of lengths is 40 cm/60 cm = 2/3. Thus the new frequency is (3/2)-times the old frequency = (3/2) × 440 Hz = 660 Hz. The new frequency has to be higher, since string is shorter.

17.2 HIGHER MODES

What happens if you gently wiggle your hand exactly twice as fast? If you understood the argument in the preceding section, you will anticipate the answer: you now make not one but two full waves for every round trip distance of the string, so that you are again re-enforcing the existing oscillation. So the next higher allowed frequency is twice the fundamental frequency, or $f_2 = 2f_1$. (Figures 17.3 and 17.4).

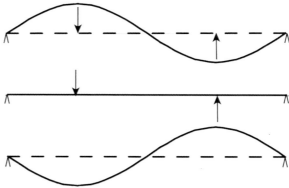

Figure 17.3. Top: oscillation of a slinky in the second mode. Bottom: photograph of a string oscillating in the second mode.

Figure 17.4. Shape of the string at different instances in time. The pictures are 1/4 period apart.

If you shake the slinky (or the string) even faster, you can produce 3 half-wavelength along it. Again all points of the slinky do simple harmonic motion.

Figure 17.5. Top: oscillation of a slinky in the third mode. Bottom: photograph of a string oscillating in the third mode.

Figure 17.6. Shape of a string at different instances in time. The drawings are 1/4 period apart.

The third mode has frequency $f_3 = 3f_1$.
The n-th mode has a frequency n-times the fundamental frequency, or:

$$f_n = nf_1 = nv/2L \qquad \textbf{(17.2)}$$

Wavelength: if you look at the picture of the string oscillating in the first mode (Figure 17.1) it looks like half of a wavelength fits into the length of the string. So we can also express the condition for the fundamental oscillation by saying that $\lambda_1/2 = L$.

Well yes, this is correct: remember that $\lambda = v/f$, so that if we substitute for f the frequency of the first mode ($f_1 = v/2L$) we get $\lambda_1 = 2L$.

For the *n-th mode* the frequency is n-times higher so that the wavelength is n-times shorter, thus *n half-wavelengths fit into the string length.*

This is summarized in the table below, and shown in Figure 17.7.

first mode (fundamental) $\quad f_1 = \dfrac{v}{2L} \qquad \lambda_1 = 2L$

second mode: $\qquad\qquad f_2 = 2f_1 \qquad \lambda_2 = \dfrac{2L}{2} = L$

third mode $\qquad\qquad\quad f_3 = 3f_1 \qquad \lambda_3 = \dfrac{2L}{3}$

fourth mode $\qquad\qquad\quad f_4 = 4f_1 \qquad \lambda_n = \dfrac{2L}{n}$

nth mode $\qquad\qquad\qquad f_n = nf_1 \qquad \lambda_n = \dfrac{2L}{n}$

Figure 17.7.

For all modes, each point of the string makes simple harmonic motion, but the frequency is now twice, three times, four times as large as for the first mode.

17.3 TRAVELING VERSUS STANDING WAVES

We should realize that these motions of the string in the various modes are quite different from the waves we have encountered before. In a "travelling wave" the wave crests moved along the water's surface - all points of the system (the water surface) moved up and down, nothing stood still. This is represented in Figure 17.8.

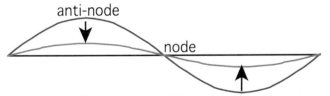

Figure 17.8. In a traveling wave, the wave crests moves along at speed v. All points move in simple harmonic motion with the same amplitude and the same frequency, but shifted in phase.

Figure 17.9. In a standing wave, all points oscillate with the same frequency but in contrast to the traveling wave all points oscillate in phase but with different amplitude. Points that never move are called *nodes*, points that have the largest amplitude are called *anti-nodes* (or "bellies").

In contrast, the wave pattern of the slinky or the string is a "standing wave" - it moves neither to the left nor to the right:: it is the sum of two waves moving in opposite directions. Here are the differences between traveling waves on a long string and standing waves on a string tied at both ends :

- In a traveling wave the crests and troughs move along the string.
- In a traveling wave all the points move up and down with the same amplitude. Different points have different phases.
- In a standing wave the amplitude is different at different points. Some points don't move at all. These places are called *nodes* of motion.
- In a standing wave points of maximum motion ("bellies" or *anti-nodes*) are halfway between the nodes. All points between two nodes move *in phase*.

17.4 THE VOICING FORMULA

To find the frequencies of the string's modes, we must know the speed of propagation v of the wave on the string. But we have learned that v depends on the tension and the mass per unit length of the string (Chapter 14). Using this result we can find the fundamental frequency of a taught string, which is tensioned by a pull F

$$\text{Fundamental frequency:} \quad f_1 = \frac{v}{2L} = \frac{1}{2L}\sqrt{\frac{F}{M/\ell}} = \frac{1}{2L}\sqrt{\frac{F}{\rho}} \qquad \textbf{(17.3)}$$

Remember that $\rho = M/\ell$ simply means the mass of one meter length of string, no matter how long the actual string is. Piano tuners and piano builders call this the "**voicing formula**" because "voicing" a piano means to put strings on it - without strings the piano has no voice!

$$\text{Frequency of higher modes:} \quad f_n = \frac{n}{2L}\sqrt{\frac{F}{\rho}} = nf_1 \qquad \textbf{(17.4)}$$

The formula 17.4 says that the frequency increases as the tension is increased, in proportion to the square root of the tension F. This means that to double the frequency the tension has to be 4-times larger. The formula says that the frequency is lower for a thicker string (larger M/ℓ), in inverse proportion to $\sqrt{(M/\ell)}$.

How does the piano builder choose the length, the tension and the mass of the string? The tension should be as high as possible (without breaking the string!) so that the oscillating string presses hard against the air in the soundboard producing a good, forceful sound volume. Bass strings need to be long and thick (massive), since L and M/ℓ are in the denominator and produce a lower f. In a large grand piano, the longest strings are about 7 feet long (2.1 m). But to make the frequency low enough the strings still need to be quite massive.

Example. A piano string is 1.2 m long and has a tension of 200 N. The tuner cut the string from a spool of wire which was labeled "200 m piano wire/mass 1kg".

What is the fundamental frequency of this string and what are the frequencies of the next two higher modes?

Answer: we substitute $M/\ell = 1.2 \text{ kg}/200 \text{ m} = 0.006 \text{ kg/m}$; $F = 200 \text{ N}$, $L = 1.2 \text{ m}$

$$f_1 = \frac{1}{2 \times 1.2}\sqrt{\frac{200}{0.006}} = 76.1 Hz$$

$$f_2 = 2 \times 76.1 Hz = 152 Hz$$

$$f_3 = 3 \times 76.1 Hz = 228 Hz$$

17.5 HOW DO MODES RELATE TO MUSIC?

When a string is bowed or plucked, many modes of oscillation occur at the same time, i.e. the actual motion of a bowed or plucked string is the *superposition of many modes*, often up to 10 or more modes.
Correspondingly, the tone produced by a 440 Hz string has in it also vibration frequencies of 2-time, 3-times, 4-times 440 Hz. Musicians call these the *higher partials* of the tone, or *overtones*. The fundamental mode is the first partial, the second mode is the second partial, etc. Since "overtone" refers to the oscillations higher than the fundamental, the first overtone is the second mode (second partial), etc.

The frequencies of the partials of a musical tone need not be exact multiples of the first mode - in fact for drums this is by no means the case, nor is it true for stiff vibrating bars (xylophone).*Partials whose frequencies are exact multiples of the fundamental are called "harmonics".*

The frequencies of the harmonics form the harmonic series:

harmonic series: f_1 $2f_1$ $3f_1$ $4f_1$ $5f_1$ *etc.*

Note that the partials have simple frequency ratios: the second partial has twice the frequency of the first, in musical terms it is the *octave* (frequency *ratio 2:1*); the fourth partial is the octave of the second partial etc. The third and second partial form a *3:2 ratio*, which in music is called a *fifth*, as mentioned in Chapter 13. We will discuss this at length in Chapter 20, where we talk about musical intervals and musical scales.

As we just mentioned, the motion of a bowed or plucked string is complicated, but can be thought of as a superposition of many modes. How much of each mode is present (its amplitude) depends on how and where the string is plucked or bowed. It may be interesting to see the motion of a string which vibrates in several modes at the same time. This can be done with a computer simulation - it may not work on all computers but try by going to the URL: http://surendranath.tripod.com/Applets/Waves/Harmonics/HarmonicsApplet.html. The mix of partials determines the tone color ("timbre") of the musical tone. The relative strength of the various partials can be represented by a bar graph, which shows the amplitude of each partial, as shown in Figure 17.10.

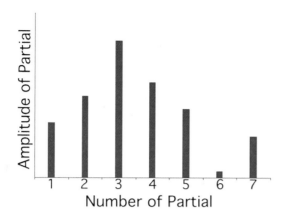

Figure 17.10. The spectrum of partials in a tone represented as a bar graph. The mix of partials determines the timbre of the musical tone. On the horizontal axis we indicate the number of the partial, while the height of the bar indicates its amplitude. A higher bar indicates a stronger partial (e.g. the 3rd partial in this graph).

17.6 DAMPING OF HIGHER PARTIALS

After plucking a string the sound slowly decays, that is, the oscillation is damped, because energy is used up in the back-and-forth bending of the string. If you listen carefully to the decaying tone of a guitar string, you may notice that the tone not only becomes softer (lower amplitude) but more and more the tone approaches a pure tone. This change in timbre is related to the fact that the *higher modes are damped faster*, because they require more bending of the string.

17.7 PLUCKED STRINGS: MISSING PARTIALS

While it is not always easy to anticipate how much of each partial is present in the tone of the instrument, you may expect that to produce a strong first mode you have to pluck the string at it's center, because this is where the first mode has an anti-node (a belly).

If you pluck the string at the exact center, is the second mode present? No - because the second mode has a node at the center, and is therefore incompatible with plucking at the center. Moving the string at the center does not excite this mode. Similarly, all higher modes with a node at the center are absent. The 2nd, 4th, 6th 8th modes are all missing, as shown in Figures 17.11 and 17.12.

Generally speaking modes that have a node at the plucking point will *not* be excited. Modes that have an anti-node at the plucking point will be strongly excited. The strength of excitation of modes depends on whether the plucking point is close to a node (weak excitation = small amplitude) or close to an anti-node (strong excitation = large amplitude). It is for this reason that a guitar string produces different sounds when plucked at different places; the closer to the bridge the strings are plucked, the more higher modes will be present, and the harsher the sound.

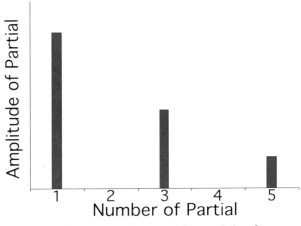

Figure 17.11. String plucked at the center: all modes with a node at the center are absent.

Figure 17.12. Spectrum of partials for a string plucked at the center shown is a bar graph..

Try to figure out which modes are missing and which ones are strongly excited, if you pluck the string at 1/3 of the string length (for instance, you pluck a 60 cm long string at a distance 20 cm from either end). From the distribution of partials in Figure 17.10 try to determine where the string was most likely plucked.

17.8 PLAYING HARMONICS

When a string vibrates in only one mode, nodes and antinodes are located in a regular pattern along its length. If you touch the string lightly at one of the nodes the vibration will continue undisturbed - if you touch it anywhere else the vibration will stop instantly.

If one plucks a guitar string near the bridge, many modes are present in the sound. If the guitarist now lightly touches the string at a point 1/4 length from the end, what modes will still sound?

Figure 17.13. Playing harmonics: we assume that the string was originally plucked near one end of the string and that a finger is lightly placed on the string 1/4 length from the end. The picture shows that the 4th, 8th, etc. partials still oscillate.
Touching the string at a given point kills all modes except the ones that have a node where the finger is placed. The modes with a node where the finger was placed survive because the string did not move at that place anyway. Figure 17.13 shows which partials survive in this case.

17.9 REAL STRINGS HAVE SOME STIFFNESS

In this section we assumed that the speed of propagation of waves on the string, v, does not depend on the wavelength (or the frequency) of the waves. This is true only if the string has no stiffness at all, and all the restoring force is provided by the tension of the string.

Real strings, however, have some stiffness (gut strings very little, steel strings quite a bit) and this will be felt as an increase in restoring force for the higher modes (where there is more bending of the string). *The stiffness of the string increases the velocity of propagation for the higher modes, and therefore slightly increases their frequency.* Musicians say that the higher modes are sharp (i.e. slightly too high in frequency).

We have learned that bass strings need to be massive, i.e. they have relatively larger diameters, which makes them stiffer. How can one make a massive metal string that is still quite pliable, that is, it has a small stiffness? One uses *overwound* strings: a second wire is wound around a central core, thereby increasing the mass without increasing the stiffness. All bass strings on the piano and the lower stings on most string instruments are overwound, but of course some stiffness remains.

Typically, for a piano string, the second mode is sharp by about 0.2%, i.e. if the fundamental is 100 Hz, the second partial is not 200 Hz but 200.4 Hz, the third partial is 300.6 Hz instead of 300 Hz, and so on. We say that the partials of the piano string are slightly anharmonic.

Chapter 18

PIPES

18.1 PRESSURE PULSE IN A PIPE

You recall from the previous chapter that a standing wave is produced by the superposition of two identical waves that travel in opposite directions In a string the opposing waves come from end reflections. Only at certain specific frequencies (or wavelengths) do these waves superimpose constructively to produce a mode or standing wave. At all other frequencies, no steady oscillation is possible because, having the wrong frequencies, after a few reflections the waves are out of step and interfere destructively.

The sound waves in a pipe are longitudinal pressure waves. An easy and quite correct way to visualize the sound wave in a pipe is to think of a slinky (to keep it from sagging we suspend it from many fine threads). If you give the slinky a push to compress it at one end (see Figure 14.5), the pulse travels along the slinky much like a pressure pulse would travel along a pipe if the air is pushed at one end of the pipe. In Figure 18.1 we show the pressure pulse traveling to the right in two different ways: as a schematic picture of a pipe shows the air density by shading, and as a graph showing the variation of air pressure along the pipe.

Figure 18.1. A pressure bump is traveling to the right in a pipe and is reflected at the open end.

When the pressure pulse arrives at the end of the pipe, it is reflected because there is a sudden change on impedance (resistance) to air motion. At the open end, the compression pulse forces the air out of the pipe into the open, and inertia keeps the air moving out until a negative pressure in the pipe sucks the air back in - so in Figure 18.1 we show the reflected pulse as a "rarefaction" (a partial vacuum) instead of a compression.

Since the wave is reflected at the end of a pipe, under the right conditions standing waves can be formed just as in the case of a string.

P.U.P.A. Gilbert and Willy Haeberli, Physics in the Arts, Part 2: Sound

18.2 REFLECTIONS IN OPEN AND CLOSED PIPES

The end of a pipe can be open, or it can be closed off with a stopper. Pipes closed at both ends are not useful as musical instruments, because there is no way for the sound to come out, or for the musician to set the air into oscillations - unless you think of a singer in a closed shower stall setting the air into oscillations!

Thus, usually by a *closed pipe we mean a pipe closed at one end, while an open pipe is open at both ends*: as shown in Figure 18.2. Unless we specify something different, it is understood that the inner hole ("bore") of the pipe is cylindrical rather e.g. a cone.

Cross section of an **open pipe**:

Cross section of a **closed pipe**:

Figure 18.2. Diagram of an "open" and "closed" pipe

In the flute the mouth piece represents an open end, even tough it is off to the side, since air is free to escape. The clarinet has a cylindrical inner bore and can be considered a closed pipe, closed off by the reeds.

It is reasonable to assume that the way the sound wave is reflected at the end of the pipe depends on whether the pipe is open or closed. At an open end, the air pressure cannot build up, since the air is free to escape, so there must be a **pressure node at the open end**. At a closed end the stopper keeps the air from moving, thus there must be a **node of motion at the closed end.** The physical conditions imposed on the air motion at the closed or the open ends are called "**boundary conditions**". For the example of the vibrating string, because the string is kept fixed at the ends, the boundary condition is that at the ends of the string there have to be two nodes of displacement.

Knowing the boundary condition makes it easy to find out what modes of oscillation are possible, because the frequency or wavelength of the sound wave in the pipe must be just right to match the boundary conditions.

BOUNDARY CONDITIONS
at open end of pipe: pressure node, motion anti-node (air escapes)
at closed end of pipe: motion node (air cannot move), pressure anti-node (pressure "belly")

It is a lot better to reason out this concept than to memorize it.

18.3 STANDING WAVES IN OPEN PIPES

The wind player learns to set the air in the instrument into periodic oscillation by making periodic puffs of air in resonance with the modes of the pipe. It may seem implausible that a musician can make for instance exactly 440 puffs of air every second when she plays the standard A on the flute - exactly in step with the resonance frequency of the flute! How this is done we will discuss later - the mechanism involves a feedback between the flute oscillations and the oscillations of the musician's air stream.

Assume that the musician has set the air in a flute (open pipe) into steady oscillations. How do we picture the motion and pressure variations in the pipe? On the left of Figure 18.3 the dark and light areas show regions of compression and decompression of the air in the pipe during one oscillation of the **fundamental mode**. In the first picture the air is compressed most at the center. Air starts to flow out both ends of the pipe until the pressure is the same as the air pressure outside - but inertia keeps the air flowing out until the pressure inside the pipe is below normal air pressure. At this point the airflow reverses and the air inside the pipe is compressed again.

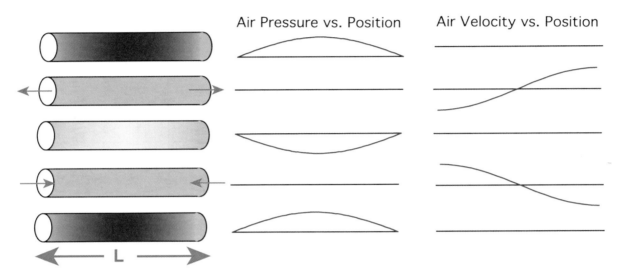

Figure 18.3. *The fundamental mode of an open pipe*. The dark and light areas of the pipe on the left indicate regions of high or low air pressure, respectively. The graphs on the right show the air pressure distribution along the pipe and the air velocity. Negative velocity means that the air moves to the left. Adjacent graphs are 1/4 of a period apart.

On the right we show the corresponding graphs for air pressure and speed of airflow. It might seem surprising that the air is still ($v = 0$ throughout the pipe) at the moment when the pressure inside the pipe is the highest. But remember the pendulum motion: when the displacement is the largest (end of the swing) the speed is zero because at that moment the direction of travel reverses, while the speed is greatest at the middle of the swing when the displacement is zero. The same happens in the open pipe here: the air speed is greatest when the air is neither compressed nor diluted.

Note that in the fundamental mode, at the center of the pipe there is a node of motion. The air at the center gets compressed and decompressed, but does not move! So the center is a node of air speed.

18.4 FUNDAMENTAL FREQUENCY OF OPEN PIPE

Just as in the case of the string, the frequency of excitation (external pushes) must correspond to the round trip travel time of the wave on the string or pipe. For a pipe, the speed of travel to use is the speed of sound in air, $v = 344$ m/s at room temperature. Thus the fundamental frequency of an open pipe of length L is:

$$\textbf{Fundamental of open pipe:} \quad f_1 = \frac{v}{2L} \qquad \textbf{(18.1)}$$

Note that the pipe diameter does not appear in the formula - the diameter of the pipe has almost no effect on the frequency of the pipe's tones. For a string, the diameter matters, but not for a pipe. You might wonder why that is, because after all for a pipe the mass of air that is oscillating is greater too when the pipe has a larger diameter. But the difference is that for a pipe of larger diameter, the push on the air by the propagating sound wave is also bigger because the air pressure acts over a larger area. The larger push offsets the larger mass. For a string too, if you increased the diameter <u>and</u> the tension the frequency would stay the same. A string of length L can produce any frequency, because we can change the mass of the string or the tension at will. But for a pipe of length L the frequency is fixed, because we cannot change the speed of sound in air! So when a string player tunes the instrument, she changes the tension, but when a flutist tunes her instrument she makes it a little shorter or longer by pulling out the mouth piece from the body of the flute.

Example*:* The flute (with all finger holes closed) plays the tone C, which has a frequency of 260 Hz. How long is the flute?

Answer: solve $f_1 = v/2L$ for L:

$$L = \frac{v}{2f_1} = \frac{344\,\frac{m}{\sec}}{2 \times 260\,\frac{1}{\sec}} = 0.66m$$

18.5 HIGHER MODES OF OPEN PIPE

Becasuse mode are sinusoidal oscillations, to find the higher modes, we must find those *sinusoidal standing waves inside* the open pipe with pressure <u>nodes at each end</u> of the pipe (boundary condition).

For the *second mode* there will now be *two half-wavelengths inside the pipe*, again with a pressure node at both ends. From Figure 18.4 we see that for the second mode there is a region of high pressure and a region of low pressure inside the pipe. Initially, air flows out of the left side and into the right side of the pipe, which is opposite to what we found for the first mode. Note that in this mode the air at the center of the pipe moves back and forth, while for the fundamental mode the air in the center never moved.

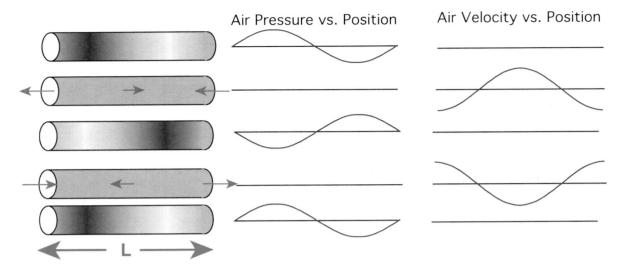

Air Pressure vs. Position   Air Velocity vs. Position

Figure 18.4. ***Second mode of an open pipe.*** The dark and light areas of the pipe on the left indicate regions of high and low air pressure. The graphs on the right show the air pressure distribution along the pipe and the air speed. Note that when the air flows out of the left end it flows into the right end, which is opposite to the fundamental mode of Figure 18.3. Adjacent graphs are 1/4 of a period apart.

For the third, fourth etc. modes, three, four etc. half-wavelengths fit inside the length of the pipe. We summarize the pressure distributions for all modes in Figure 18.5, where we mark the pressure nodes (places of constant pressure) with a green arrow and the anti-nodes with a red arrow. Remember that the air velocity is opposite: *the air velocity is largest where the pressure has a node, and smallest where the pressure has an anti-node*.

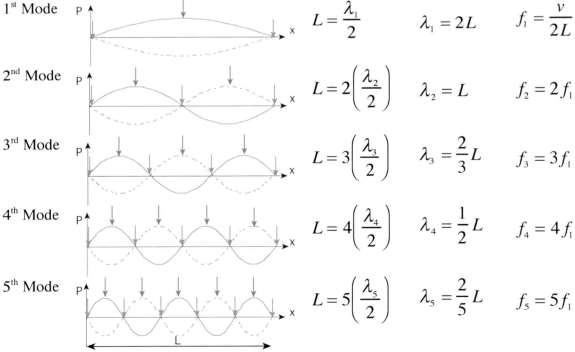

1st Mode

$$L = \frac{\lambda_1}{2} \qquad \lambda_1 = 2L \qquad f_1 = \frac{v}{2L}$$

2nd Mode

$$L = 2\left(\frac{\lambda_2}{2}\right) \qquad \lambda_2 = L \qquad f_2 = 2f_1$$

3rd Mode

$$L = 3\left(\frac{\lambda_3}{2}\right) \qquad \lambda_3 = \frac{2}{3}L \qquad f_3 = 3f_1$$

4th Mode

$$L = 4\left(\frac{\lambda_4}{2}\right) \qquad \lambda_4 = \frac{1}{2}L \qquad f_4 = 4f_1$$

5th Mode

$$L = 5\left(\frac{\lambda_5}{2}\right) \qquad \lambda_5 = \frac{2}{5}L \qquad f_5 = 5f_1$$

Figure 18.5. Each graph is a graph of pressure vs. position inside the pipe. Red arrows show pressure anti-nodes, and green arrows show pressure nodes. Notice that in odd harmonics there is a pressure anti-node at the center of the pipe, while in even harmonics there is a node at the center of the pipe.

As you can see from Figure 18.5, the higher the mode the shorter the wavelength, and therefore the higher the frequency. But all frequencies are multiples of the first mode frequency: they are *harmonic*. This is very important from the musical standpoint. The fact that the modes of pipes and strings are harmonic is responsible for the pleasing sound of the instruments.

18.6 FUNDAMENTAL MODE OF CLOSED PIPE

As mentioned in section 18.2, for a *closed pipe* (closed at one end) there must be a node of pressure at the open end but an *anti-node of pressure at the closed end*. At the closed end the air pushes against the stopper, so that the air pressure varies at the closed end, while at the open end the air is free to escape and the pressure is constant. The pressure and air velocity graphs for a closed pipe are shown in Figure 18.6.

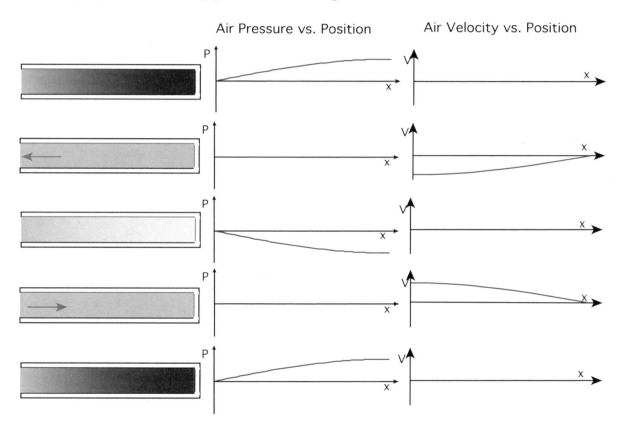

Figure 18.6. **Fundamental mode of closed pipe.** The dark and light areas of the pipe on the left indicate regions of high or low air pressure. The graphs on the right show the air pressure distribution along the pipe and the air velocity. Negative velocity means that the air moves to the left. Adjacent graphs are 1/4 of a period apart.

Fundamental frequency of closed pipe

You may have noticed in Figure 18.6 that for the fundamental oscillations of the closed pipe 1/4 of a wavelength fits inside the pipe of length L, while for the open pipe the wavelength was half as long, so that 1/2 a wavelength fits. In other words, considering two pipes of the same length: the closed pipe has twice the wavelength of the open pipe. Longer wavelength means lower frequency: *a closed pipe has half the frequency of an open pipe of the same length.*

Expressed as an equation:

Fundamental of closed pipe: $\qquad f_1 = \dfrac{v}{4L}$ **(18.2)**

A particularly attentive reader may argue that if we again were to use the argument that the round trip time of a pulse should equal the period of the fundamental, we would get the same answer for the period no matter whether the pipe is open or closed. This apparently is not the case. Why not? The answer is that at the closed end a positive pulse is reflected as a negative one, as shown in Figure 18.1. Thus a belly of the sine wave is reflected at the closed end as a through, meaning that the phase of the wave jumps by 180°. So we loose a half-wavelength in the reflection at the closed end.

Example. A 2 m long organ pipe is closed at one end. Find the fundamental frequency.

Answer: $f_1 = \dfrac{v}{4L} = \dfrac{344\,m/s}{4 \times 2m} = 43Hz.$

On a pipe organ, the pipes of lowest frequency are often closed pipes. To get the same 43 Hz fundamental from an open pipe requires a pipe twice as long, i.e. 4 m or about 12 feet.

18.7 HIGHER MODES OF CLOSED PIPE

For the next higher mode, we must again be sure that there is a pressure node at the open end and a pressure anti-node (belly) at the closed end. The picture for the second mode is shown in Figure 18.7.

So what is the frequency f_2 of the second mode? There are now **three** (1/4) waves inside instead of just one, so the wavelength is three times shorter and the frequency **three-times higher** than the fundamental: $\mathbf{f_2 = 3f_1}$. In contrast to strings and open pipes, the next mode does **not** have twice the frequency!

Air Pressure vs. Position   Air Velocity vs. Position

Figure 18.7. A closed pipe oscillates in the second mode. The pressure has a node at the open end and an anti-node at the closed end. The air velocity has an anti-node at the open end (air rushes in and out) but a node at the closed end because the stopper keeps the air at the closed end from moving. The length of the pipe is L=(3/4)λ.

Below we show the pressure distribution for the first five modes of the closed pipe. To always have a anti-node (belly) at the closed end, we must use an **odd multiple of λ/4** inside the pipe: namely 1, 3, 5, 7, etc. Correspondingly, the frequencies of the higher modes are 1, 3, 5, 7, 9-times the fundamental frequency., and not 1, 2, 3-times the fundamental!

$$\textbf{closed pipe:} \qquad f_n = (2n-1)\times\frac{v}{4L} \text{ where n=1, 2, 3,.....} \qquad \textbf{(18.3)}$$

Example. Let us find the first three mode frequencies of a 2 m long closed pipe.
Answer: we found above that the fundamental of a 2 m long closed pipe is 43 Hz. The next two modes have frequencies of 3 × 43 Hz = 129 Hz and 5 × 43 Hz = 215 Hz.

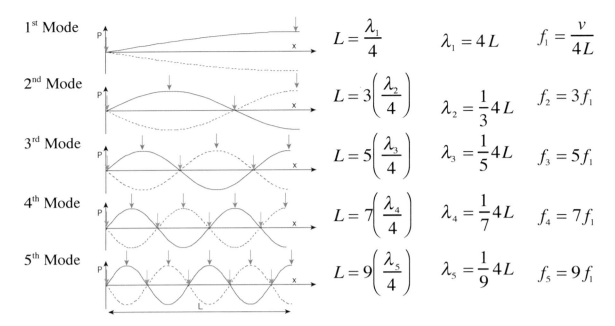

Figure 18.8. First five modes of closed pipe. The graphs show pressure vs. position inside the pipe. Red arrows show anti-nodes of pressure, and green arrows show nodes of pressure.

18.8 PLAYING TUNES ON WIND INSTRUMENTS: FINGERHOLES AND "OVERBLOWING"

Just as for the string, in wind instruments the player does not excite only a single mode, but many modes at the same time. *The sound from an open pipe is the sum (superposition) of many sinusoidal oscillations of frequency f_1 and multiples of f_1, whereas for the closed pipe only the odd multiples of f_1 are present.*

String players can easily change the fundamental frequency simply by putting a finger on the fingerboard to make the string shorter. Wind players have to make their pipe shorter or longer by opening different finger holes where the air can escape and make a pressure node somewhere along the pipe, or by making the pipe longer or shorter with valves (trumpet) or a slide (trombone).

But it would be difficult to cover a very large range in frequency this way - every octave requires making the pipe half as long so after three octaves it is 1/8 the original length and there is almost nothing left - not to speak of the tremendous number of finger holes the musician would have to open and close. Another way to produce frequency changes in wind instruments is to *"overblow"* to the next mode: the wind player learns to blow in such a way that the excitation frequency of an open pipe jumps to the second mode, so the full length of the pipe plays $2f_1$, that is, the octave. Or even three or four times the fundamental. One striking application of this is the bugle, which is a trumpet without any valves - the bugle can nevertheless play a sequence of tones: f, 2f, 3f, 4f etc., called the harmonic series or the harmonic scale. By what magic the musician can at will excite different modes we will discuss in Chapter 21.

18.9 OTHER SHAPES

Except for the flute and some organ pipes, wind instruments are not usually simple cylinders. Brass instruments have flares attached to the end, not only to improve sound radiation but also to shift the resonance frequencies. Some instruments (oboe, bassoon) have a conical rather than a cylindrical opening (see Chapter 21).

The math for figuring out the modes of these instruments is complicated, but we want to mention one interesting fact. The modes of a *conical pipe* are for some reason mathematically identical those for an <u>open</u> cylindrical pipe of the same length. Of course for a conical pipe we imagine a mouthpiece attached to the tip of the cone for the musician to blow, so that the air inside is set into oscillation.

18.10 ACOUSTIC LENGTH

In connection with equation 18.1 we mentioned that the diameter of the pipe does not appear in the frequency formula, i.e. the frequency is the same no matter how large a diameter the pipe has. But this is an oversimplification: in actual fact, the pipe oscillations extend beyond the physical end of the pipe by some fraction of the pipe diameter. The effect of this extra length is to slightly lower the frequencies of all the modes. The "acoustic length" of a pipe is the value of L in Equation 18.1 that corresponds to the **actual** frequency of the pipe. The amount by which the acoustic length exceeds the physical length is about 0.6-times the radius of the pipe.

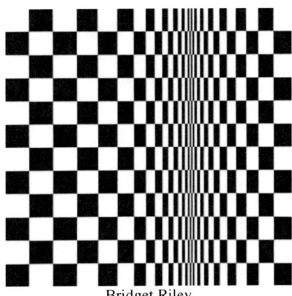

Bridget Riley
"Movement in Squares" (1961)

19.1 THE FOURIER THEOREM

When we looked at modes of a plucked string, we represented the intensity of the various modes by a bar graph, which showed the amplitude of all the modes that make up the complex vibrations of the string. A bar graph of course is much easier to draw than the complex wave which is the superposition of the various sine curves - you may remember from the examples in Chapter 13 how tedious it is to draw the actual wave shape when two oscillations are superimposed, not to mention three or four!

Looking at a wave shape, for instance of the human voice on an oscilloscope, may be fun but very hard to interpret, or to identify which feature of the wave shape is different when you compare the voice of an opera star to your own voice. A young singer might decide to improve her voice by trying to make her oscilloscope trace look similar to the oscilloscope trace from the recording of a famous star, but this would be futile because the oscillations are so complex. In addition, changing the phase (the timing) between oscillations changes the wave shape but not the tone we hear, which is another indication that the wave shape is not all that important. Rather it is the intensities of the partials – and not their relative timing - that determine the tone quality.

Instead of dealing with the complicated wave shapes that voices and instruments produce, *Fourier analysis* will allows us to describe a steady tone of any complexity by just a few numbers, or by a simple graph.

Baron Jean Baptiste Joseph Fourier, a French mathematician (1768 - 1830), presented a mathematical proof of the "Fourier theorem". This theorem has three parts:

1. Any periodic wave of period T can be represented as a sum of sine waves with appropriate frequencies, phases and amplitudes.

2. Only certain frequencies are necessary, namely those that are multiples of the fundamental frequency f = 1/T.

3. There is only one combination of sine waves that produces the given complex wave.

In other words, the Fourier theorem states that:
Any periodic function of period T can be expressed as a sum of sine curves whose frequencies are multiples of f = 1/T and whose amplitude and phase are suitably chosen.

For example, if a singer's voice makes a very complex oscillation of some repetition frequency f, this oscillation can be represented by the sum of *Fourier components*, namely sinusoidal oscillations of frequencies f, 2f, 3f, etc. Of course for a string or a pipe we are not surprised that this is possible, because we have learned from the physics of their oscillations that the tone they produce is a superposition of modes of frequency f, 2f, 3f, etc. Musicians call the Fourier components the *partials* of the tone.

Why are the frequencies of the Fourier components exact multiples of the fundamental frequency? If the fundamental oscillation has a period T, the periods of the higher oscillations must fit into T once, or twice, or three times, etc. If they did not they would spoil the periodicity. Thus we readily understand that the frequencies of the Fourier components must be multiples of the fundamental frequency.

19.2 SOUND SPECTRUM

The Fourier spectrum of a musical tone (also called the **sound spectrum**) is a graph that represents the amplitude of the various partials of a complex tone vs. the frequency of the partial. Often, if there are lots of partials, we may show the number of each partial instead of the actual frequency, but we still need to specify the frequency of the fundamental.

How does one find the Fourier components (partials) of a complex tone? There are two ways: one can record the exact wave shape and use a mathematical procedure involving integral calculus to find the amplitudes of the partials. For studies in acoustics, a much more convenient and elegant approach is to use a "Fourier analyzer" which carries out this calculation and immediately displays the spectrum of sound picked up by a microphone on a computer screen. Any present day desktop computer with a sound card and either an internal or external microphone can be used as a Fourier analyzer. The necessary programs are now readily available for free or for a small fee on the internet. Simply search for "Fourier analyzer" – you will have great fun exploring your voice or a musical instrument.

To get used to the concepts, we first show below graphs of some periodic oscillations and the corresponding Fourier spectrum. We first deal with calculated wave shapes (Figures 19.1 to 19.4) and later show examples of actual measurements.

1. A pure tone (e.g. tuning fork): the left of Figure 19.1 shows the sound pressure variations. The right shows the corresponding Fourier spectrum. It is obviously a lot easier and more instructive to draw the Fourier spectrum than the actual air pressure variations. The graph tells us that there is one sine wave present of frequency 300 Hz and amplitude 2 units.

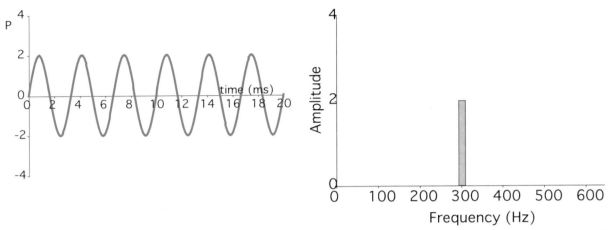

Figure 19.1. Pressure variations and corresponding sound spectrum of a pure tone of frequency 300 Hz and amplitude 2 units.

2. A pure tone of different frequency and amplitude:

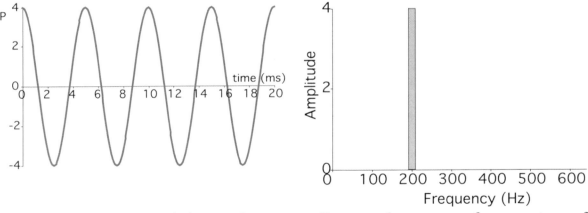

Figure 19.2. Pressure variations and corresponding sound spectrum of a pure tone of frequency 200 Hz and amplitude 4 units.

3. A complex tone consisting of two partials: In Figure 19.2 we have added the above two oscillations to produce a more complex tone. The period of the resulting oscillation is 10 ms, which corresponds to a 100 Hz frequency.

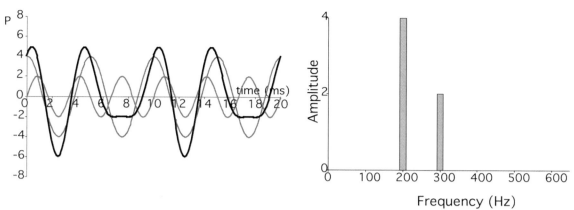

Figure 19.3. Pressure variations and corresponding sound spectrum when the two tones of Figures 19.1 and 19.2 are played at the same time.

Now it is even more obvious that the Fourier spectrum is an elegant way to see what is happening - if you only were given the complex oscillation (black curve in Figure 19.3) you would have a hard time figuring out its components, while the Fourier spectrum immediately tells you what partials made up the black curve

The missing fundamental: in the above example, the complex oscillation clearly has a frequency of 100 Hz, i.e. it repeats itself every 10 ms, or 100-times per second. As we explained earlier, this frequency is the largest common divisor of 200 Hz and 300 Hz. But what is the amplitude of the fundamental in the Fourier spectrum? The fundamental has amplitude 0 (zero), i.e. there is no fundamental present - yet 100 Hz is the fundamental in the sense that the repetition frequency is 100 Hz. We refer to this as the "missing fundamental".

It is interesting to ask ourselves: what frequency do we actually hear in this case? If you have several partials, but the fundamental is not actually present in the complex oscillation, the pitch still sounds like the fundamental that is not there! This is one reason why a small radio on your nightstand that has virtually no bass response still can reproduce bass sounds to some extent, because the higher partials of the actual bass suggest to the ear-brain system the presence of the fundamental, even if the speaker in the radio does not reproduce it.

4. Change in phase affects wave shape but not Fourier spectrum. Drawing the spectrum instead of the wave shape is a lot easier and more instructive, but here is a possible problem: When you change the timing (phase) between the oscillations, the wave shape changes considerably, as shown in the next graph. But the spectrum graph stays the same because the graph only says how much of each oscillation there is, not how they are timed. So how can one say that the spectrum graph represents the real sound?

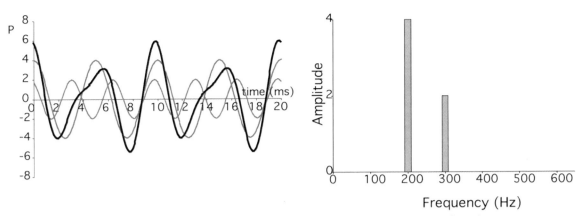

Figure 19.4. Pressure variations and corresponding sound spectrum for the same two tones of Figure 19.3, but shifted in relative timing (phase). The pressure variations are different but the sound spectrum and the perceived sound stays the same.

It is indeed sufficient to show only the amplitude but not the phase of the partials because we cannot hear differences in phase between the partials! Two oscillations (e.g. left hand side of Figures 19.3 and 19.4) can have rather different wave forms yet sound exactly the same. We say that *the ear is insensitive to the phase of the partials*. Stated another way: the human hearing apparatus is a Fourier analyzer that takes apart the complex wave shape into its components (partials) without paying attention to the detailed wave shape. This is rather surprising in view of the coding of sound in nerve pulses along the auditory nerves, the distribution of which in fact is sensitive to phase - so it is only later, in the brain, that the phase information is ignored.

This raises another question: when we listen to a complex tone from a musical instrument, do we hear that the tone is made up of partials? The answer is no. We integrate all the partials into one perception of one tone, with one frequency. In our perception, all the partials fuse into a single tone, which we identify as a flute, a violin etc. This process – *fusion of partials* – happens in the brain and is poorly understood. Some musicians claim that they can hear partials but most of us don't.

5. The sound spectrum of musical instruments
Below we show the sound spectra for two instruments: an alto Saxophone and a Clarinet, as shown in Figure 19.5. This illustrates that real instruments often have many partials. A violin would have even more. Secondly it illustrates that the clarinet has strong odd and weak even partials (at least for the first few partials) similar to a closed pipe, while the saxophone has strong even and odd partials. More about that when we talk about instruments in Chapter 21.

Figure 19.5. Sound spectrum of an Alto Saxophone playing a tone with frequency 233 Hz, and a clarinet playing a frequency of 134 Hz. The clarinet has strong *odd partials*, the saxophone has odd and even partials of similar strength.

In Figure 19.6 we show the spectrum of two organ pipes playing the same tone C_4, which has a frequency of 262 Hz. The names of organ pipes are often derived from German names - the name "gedackt" pipe comes from "gedeckt" or covered, i.e. closed pipe (left graph). It shows odd harmonics only. In contrast, the Spitzflöte ("spitz" = point, i.e. a pointed, conical pipe), has odd and even harmonics. The two pipes have the **same pitch but very different timbre**: the Spitzflöte sounds very nasal while the closed pipe has a smoother, rounded tone.

Figure 19.6. sound spectrum of a closed organ pipe (left) and a conical organ pipe (right), both playing a frequency of 262 Hz. Note that the closed pipe has only odd partials.

19.3 FOURIER ANALYZER (SOUND ANALYZER)

Almost any computer with a sound card and a microphone can be used to record and to graph almost instantaneously the sound spectrum of a tone played or sung into the microphone. Such displays are now used in voice training: experienced singers know how to sing a vowel sound in such a way that all partials are immediately present at the beginning of the tone, while for less experienced singers the higher partials come into play only slowly. By watching the spectrum on the computer screen, the singer learns how to avoid this undesirable "blooming" of the partials.

Two examples of sound spectra acquired with a computer are shown below.

In Figure 19.7 one of the authors was singing "aah" as in f<u>a</u>ther" at about 220 Hz. He actually tried to match the frequency of a 440 Hz tuning fork, but sang the tone an octave too low. It is interesting that even good singers when they try to match the tone of an instrument often find that they sing in a different octave.
The top panel of the computer screen reproduced in Figure 19.7 shows the air pressure variations picked up by a microphone. Note that the wave is periodic, but quite complicated. The time scale shows that the period is approximately 5 ms. The panel below shows that the sound spectrum contains partials up to the 18th partial. There are stronger groups in the region up to about 1000 Hz, and then again around 3000 Hz. These regions are called the **formants** of the vowel "aah" in Willy's voice and are the result of broad resonances in the throat and oral cavity.

Figure 19.7. Fourier analysis done on a computer. The example shows Willy's voice when he sings a sustained "aah" sound, at 220 Hz. The computer displays the sound pressure variations detected by a microphone (top part of the computer screen) and the sound spectrum (bottom part of the screen).

Figure 19.8. Sound spectrum of a violin playing a frequency of 350 Hz. Note that the 10th partial is stronger than the fundamental

Notice the difference between Willy's voice and a violin, by comparing Figures 19.7 and 19.8.

19.4 FOURIER SYNTHESIS

If any periodic oscillation can be represented as a sum of sine waves, it must be possible to artificially synthesize any periodic oscillation by adding sine oscillations. Accordingly, a violin tone or Willy singing "aah" can be synthesized by adding many sine waves. Electronic synthesizers produce musical tones resembling real instruments in this way - but a more interesting use of such synthesizers is to produce new and different sounds.

As an illustration of Fourier synthesis, let's see if we can synthesize a square wave as the black one in Figure 19.9 with a combination of sine curves. We start by making a fundamental sinusoidal oscillation of frequency f that resembles the square wave as much as possible (red sine wave in Figure 19.9). The next higher oscillation will have twice the frequency of the fundamental (cyan sine). But you can see that adding such an oscillation is a bad idea, because the square wave is mirror-symmetric left and right of the vertical dashed lines, while the oscillation with frequency $2f$ is not symmetric. Thus we should make the amplitude of the 2f oscillation zero. Adding 3f, on the other hand, really helps (Figure 19.10). Adding a bit of 3f to the fundamental already makes the sum of these two a lot closer to the square (purple line in Figure 19.10).

In Figure 19.11 we show that the sum of 4 harmonics (f, 3f, 5f, 7f) is getting closer to the square wave, but note that the steep vertical sides of the square wave cannot really be made up of low-frequency sine waves. To produce a vertical step would require infinitely high partials. Finally, Figure 19.12 shows that 51 partials are a pretty good representation of the square wave.

It is generally true that to represent a very rapid rise or fall of the curve requires very high frequencies.

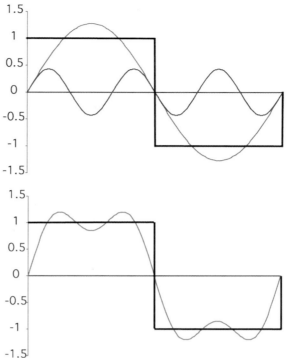

Figure 19.9. Top: a square wave (black curve) is approximated by the red sine wave.
Bottom: the square wave is mirror-symmetric about the dashed vertical lines Adding an oscillation of twice the frequency would destroy the symmetry.

Figure 19.10. Top: here we added an oscillation of frequency 3f to the fundamental of frequency f.
Bottom: the sum of these two shows that we are already getting closer to representing a square wave (lower graph).

Figure 19.11. Adding more waves of the proper frequency (odd multiples only) and the right amplitude makes the result closer and closer to a square wave.

Figure 19.12. The sum of 51 odd harmonics, shows a very good approximation of a square wave.

19.5 WHY CAN'T WE SYNTHESIZE A STRADIVARI?

Can we synthesize the tones musicians play? Indeed it is easy enough to synthesize a perfectly steady, perfectly periodic oscillation, but this is not what a musician aspires to play. In the first place, a musical tone is not a start-play-stop oscillation. Even though we are not consciously aware of it, the tone builds up over a fraction of a second and at the end it does not stop abruptly, but takes a certain time to decay. The characteristic attack and decay of a tone are both important. If a steady tone of an instrument is recorded and then replayed with the beginning and the end of the tone cut off abruptly, the instrument looses so much of it's character that in some cases it becomes nearly unrecognizable. For a synthesizer to reproduce the attack and decay one must keep in mind that each of the approximately 20 partials has a different rise and decay time, which adds a lot of complexity.

But a more fundamental problem is that real musical tones are not completely periodic. Take a violinist as an example: as she plays a tone, she has several parameters she can vary with time during the tone: speed of bow, pressure on bow, distance of bow from bridge, finger position (vibrato). But each time she plays the same tone again, she will choose slightly different parameters and it is these choices that distinguish artists from one another even if they were to play the same Stradivari! Thus no – there is no point trying to synthesize the beautiful, lively, varied tones musicians play, be it on a Stradivari or any other instrument. The main purpose of synthesizers then is to produce interesting new sounds, different from actual instruments.

Chapter 20

MUSICAL SCALES

A musical scale is a sequence of frequencies, most often arranged in ascending order. A scale normally is contained within one octave, i.e. within a frequency range of 2:1. The tones of the scale are also referred to as the notes of the scale. The term "note" is used either to denote the symbols that represent the tones of the scale on paper or the tones themselves.

Why do we need a scale? To make music in an organized fashion one needs some set of frequencies that people (more or less) adhere to, at least within a given culture. Otherwise, how would one tune a piano and boreholes in the body of wind instruments etc etc? And how would an entire chorus sing "in tune"?

We could simply describe the Western scale by a table of frequencies and could say that's the way the scale developed over the centuries. But physics tries to understand why it is that way – why does the piano keyboard end up with the funny arrangement of black and white keys where a black key is missing between some of the white keys? It is clear that a scale evolves over centuries, based on the esthetic judgment of generations of musicians, but here we seek to find the reason in particular for the development of the scale used in most Western music. We want to show that quite likely the scale as we know it is a direct result of the desire to make a scale that allows for harmonious frequency ratios. Some of it will involve some tedious calculations, but the reader who is only interested in the end result can skip over it.

20.1 MUSICAL INTERVALS

When you sing a "tune" your voice produces a sequence of musical tones of different pitch, that is, a sequence of sounds with different frequencies. As a child, if you grew up in the US or in Europe, you may have learned the song "Are you sleeping, are you sleeping, brother John?" (or "Frère Jacques, frère Jacques, dormez vous?", or "Fra Martino, campanaro, dormi tu?", or "Bruder Jakob, BriderJakob, schläfst du noch".)

	Are	you	sleep-	-ing
	150 Hz	**168.7** Hz	**187.5** Hz	**150** Hz

But this is not the only way one can sing this tune: A woman may sing it at a higher pitch:

Are	you	sleep-	-ing
200 Hz	**225** Hz	**250** Hz	**200** Hz

or a man with a low voice may sing it like this:

Are	you	sleep-	-ing
100 Hz	**112.5** Hz	**125** Hz	**100** Hz

These different sequences of tones to us are "the same tune" that is, the same sequence of musical intervals. The question is: what about these frequencies is the **same** for all three sequences of numbers above?

The answer is simple. *the ratios of the various frequencies are the same*. The frequency of the second tone divided by that of the first in each case is the same, namely 9/8. Let us check:

$$\frac{112.5}{100} = 1.125 = \frac{9}{8}; \qquad \frac{168.75}{150} = 1.125 = \frac{9}{8}; \qquad \frac{225}{200} = 1.125 = \frac{9}{8}$$

The ratio between the next two tones, "you" and "sleep-" is also the same for all three renditions of the tune.

We must therefore conclude that what sounds to us like the same musical interval, in physical terms means the same frequency ratio. Note that the frequency differences is Hz are quite different, it is the **ratio** that matters:

A musical interval is characterized by the ratio of the two frequencies. Intervals sound the same if they have the same frequency ratios.

This means that if you divide (or multiply) all frequencies of a tune by the same number, it will again be the same tune. On some old record players you can change the speed of the turntable. If you change the speed, the music of course is faster or slower, but it is exactly the same tune in a higher or lower pitch - all the frequencies are multiplied by the same number.

Examples of musical intervals (the origin of the **names** is explained later):

A **2:1** ratio is called the *octave*: the interval from 200 Hz to 400 Hz sounds the same as the interval from 300 Hz to 600 Hz.

A **3:2** ratio is called a *fifth*: the interval from 200 Hz to 300 Hz sounds the same as the interval from 300 Hz to 450 Hz.

A **5:4** ratio is called a *major third*: in the example above ("Are you sleeping..."), the third tone ("sleep-") is a major third above the first ("Are").

A **6:5** ratio is called a _minor third_.

Note that a 3:2 ratio (3/2 = 1.5)) is the same interval as the inverse 2:3 ratio (2/3 = 0.666).
The only difference is that in one case we "go up a fifth", in the other case we "go down a fifth".

Example: what is the frequency of a tone a fifth above A = 440 Hz? A tone a fifth below?
Answer: going up means to multiply by a 3/2: Thus the frequency becomes 440 × (3/2) = 660 Hz.
Going down means either to **divide by 3/2 or multiply by 2/3**. Multiplication by some fraction like 2/3 is the same as division by the reciprocal 3/2:

$$440 \times \frac{2}{3} = \frac{440}{\frac{3}{2}} = 293.3 \text{Hz}$$

20.2 CONSONANCE (HARMONY): SIMPLE NUMBER RATIOS

In all musically developed cultures tunes are played with a number of tones that have well-defined intervals. These are the intervals of the _musical scale_. The tuning of the piano or the placement of frets on the guitar are chosen to correspond to those particular intervals. But what exactly are those intervals and how were they chosen to start with
The establishment of a musical scale is a cultural development that extends over centuries and involves subjective preferences. However, one finds that all cultures seem to have a preference for certain musical intervals (frequency ratios), like the octave, the fifth and other simple-number ratios. The preference for simple-number ratios quite likely is related to the fact that in most music two or more tones are heard simultaneously. Compositions in which two or more tones are sounded simultaneously are called "polyphonic" (from the Greek: Poli = many; phonoes = sound). Not all tones played in combination simultaneously sound pleasing to the ear. Some intervals are "rough" or "dissonant", while other intervals are particularly pleasing, harmonious or "consonant". But what combinations of frequencies f_1, f_2, ... are in harmony and why? A plausible explanation is the following:

Simple-number frequency ratios are consonant ("in harmony") because the various partials of the two tones being played simultaneously either coincide exactly or are different enough to avoid beats or roughness.

"_Roughness_" comes about when two frequencies are near one another but not so close that one discerns the beats individually (e.g. 400 Hz and 410 Hz). For example, two tones a fifth apart, such as 200 Hz and 300 Hz are consonant, because the partials of one (200, 400, 600, 800, 1000, 1200 Hz) and of the other (300, 600, 900, 1200 Hz) have a particular relationship: some partials coincide exactly (here 600, 1200 Hz) while the others are so different that they produce no roughness. This of course would not be the case for a detuned fifth, such as 200 Hz and 305 Hz, since now the third partial of 200 Hz (600 Hz) clashes with the second partial of 305 Hz (610 Hz).

An extreme case of consonance (harmony) is the octave (e.g. 200 Hz and 400 Hz), where all the partials of the higher tone (400, 800, 1200 Hz) are already present as partials of the lower tone (200, **400**, 600, **800**...), so the higher tone has nothing new to contribute. This then explains why two tones an octave apart sound so alike and in music are given the same name. The tone of 440 Hz and the octave 880 Hz are both called A, but are distinguished by a subscript: A at 440 Hz is called A_4 while the octave above, A at 880 Hz, is called A_5. In this way, A_0 is near the bottom of the audible range.

To repeat: tones with simple number ratios of frequencies are consonant because of absence of beats or roughness between their higher partials. We said "simple number ratios"– but how simple do they need to be? And experiment shows that 6:5 is perceived as consonant (in earlier times it was considered dissonant!) while already 9/8 is not and 16:15 is very dissonant!

20.3 THE MAJOR TRIAD

A "triad" is a set of three tones, often played together as a "chord". If you learned to sing or to play an instrument you will know the names of the tones of the (western) musical scale, like do-re-mi-fa-sol etc. or C-D-E-F-G- etc.

If you skip every other note starting on C you get a sequence C - E - G, which is called the *major triad*. What is special about it is the simple relationship between the three frequencies:

Frequency ratios of the major triad:	**4**	**5**	**6**
divide by 4	1	$\dfrac{5}{4}$	$\dfrac{3}{2}$

To get the frequency in Hz, multiply the three numbers by any number you like. For instance:

Multiply 4, 5, 6 by 50:	**200 Hz**	**250 Hz**	**300 Hz**
The tones in the usual notation are:	**C**	**E**	**G**

In reality, C is tuned to 264 Hz, but to keep things simple we use 200 Hz for the first tone. Remember, all that matters to music is the ratio of frequencies, not the frequency itself. If you used 264 Hz for C, E becomes $264 \times (5/4) = 330$ Hz. The first interval of the major triad, 5/4, is what we called a *major third*. The second interval, 6/5, is a *minor third*. The overall interval (C to G), 6/4 = 3/2 is a *fifth*.

What is special about the major triad is that it contains three simple-number ratios and thus sounds especially pleasing when the three tones are played either at the same time or in sequence.

20.4 CONSTRUCTING A SCALE: THE JUST SCALE

Because people like harmony, a scale should permit a lot of harmonious intervals. With this idea in mind, we can construct a scale step by step, as described here. Through the physics of sound, therefore, we can explain why the currently used scale most likely came about .

To play music we need more than just three tones, we need a "scale" with more tones. If C is the beginning of the scale, then the octave above is C again, but what frequencies do we put in between? Greek philosophers placed great value on numbers, and since they knew from experiments with vibrating strings that a fifth (3/2 ratio) is a perfect harmony, one idea was to keep going up in frequency in fifths, that is, repeated multiplication by 3/2. Already the second fifth goes beyond the octave, since $(3/2) \times (3/2) = 9/4$ is larger than 2. But they could go down an octave, which meant to multiply by 1/2, which makes the new frequency 9/8. This is nice when played after C, as in the very beginning of the tune "Are you sleeping..." The scale based on fifths is called the Pythagorean scale. You can calculate the ratios for this scale yourself with a calculator or a spreadsheet on a computer. But the fly in the ointment is that multiplication by 3/2 (going up in fifths) followed by multiplication by 1/2 (going down in octaves) never makes a frequency that is double of the first frequency, so we cannot even make a perfect octave. What a shame!

Here we present another way to make a scale. Since the major triad is a perfect harmony, we build the scale by **adding more tones forming major triads**. Before we write down any numbers, here is the plan. The end result will be a sequence of frequencies for the tones of the scale:

<u>C</u>   D   <u>E</u>   F   <u>G</u>   A   B   <u>C'</u>   D'   E

Let us start with the triad we already know, called major triad C - E - G, and add the octave of C (we can call it **C'** to distinguish it from the lower C). The plan is link two further triads to the first one:
One triad **starting on G**: in musical notation, these will be tones called **G - B - D**. One triad **ending on C'**: in musical notation, these will be tones **F - A - C'**.
In other words, we will choose the frequencies such that when we skip every other tone of the scale we get a perfect triad.
So the other two triads are reported in green and in blue:

C   D   E   F   <u>G</u>   A   <u>B</u>   C'   <u>D'</u>   E

C   D   E   <u>F</u>   G   <u>A</u>   B   <u>C'</u>   D'   E

Now we need to calculate the required frequencies. Again, we call the first frequency C=1.
Figure 20.1 shows the three triads separately. To get the numbers for the green triad, starting on G, since we already fixed G at G = 3/2, we now must multiply by (5/4) to get the second tone: B

= (3/2) × (5/4) = 15/8. The third tone must be (3/2) above the first, thus D' = (3/2) × (3/2) = 9/4. The D an octave below will be (1/2) × (9/2) = 9/8.

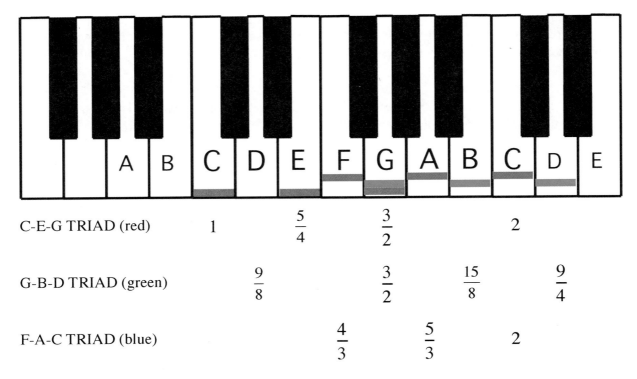

C-E-G TRIAD (red)	1		$\frac{5}{4}$	$\frac{3}{2}$			2		
G-B-D TRIAD (green)		$\frac{9}{8}$		$\frac{3}{2}$		$\frac{15}{8}$		$\frac{9}{4}$	
F-A-C TRIAD (blue)					$\frac{4}{3}$	$\frac{5}{3}$		2	

Figure 20.1. The three major triads of the just scale are indicated on a piano keyboards.

To calculate the frequencies of the last triad, we must end the triad on C' = 2. So the first tone, a fifth below, must have a frequency 2 × (2/3) = 4/3 (remember to go down a fifth we divide by 3/2 or multiply by the reciprocal, 2/3). Finally, from F = 4/3 we go up a major third (5/4) and get 5/4 × 4/3 = 20/12 = 5/3.

Here is the final result of our labors. We have constructed a scale from three beautiful, perfect triads. This sequence of frequency ratios relative to the first tone of frequency 1 is called the *just major scale*. If the first tone of the scale is chosen to be C , the scale is called the **just C-major scale**.

To find actual frequencies in Hz, again you would need to multiply all those ratios by whatever number of Hz the first tone has. In the table below we called the first tone C, but the first tone of the scale need not be tuned to the actual C of the piano - you could start the scale on a lower or higher pitch.

Name	DO	RE	MI	FA	SOL	LA	TI	DO
	C_4	D	E	F	G	A	B	C_5
Ratio to C_4	1	9/8	5/4	4/3	3/2	5/3	15/8	2
Decimal equivalent	1	1.125	1.25	1.33	1.5	1.66	1.875	2

Here we have used the notation C_4, C_5 for tones an octave apart. The above table is valid for any other octave.

Extension to lower and higher octaves: frequencies an octave above those shown in the table are twice as large, frequencies two octaves above are $2 \times 2 = 4$-times as large etc. Frequencies an octave below are half the frequencies the frequency etc.
How do we use these numbers?

Examples:
1. If C_4 has a frequency of 264 Hz what is the frequency of F_4?
    Answer: the ratio between F_4 and C_4 is 4/3.
    $$\frac{4}{3} \times 264\,Hz = 352\,Hz.$$

2. If A_4 is 440 Hz what is the frequency of E_4?
    Answer: we need to find the frequency ratio between the two tones:
    Let us calculate the interval E-A:
    $$\frac{E}{A} = \frac{5/4}{5/3} = \frac{5}{4} \times \frac{3}{5} = \frac{3}{4}$$

3. If G_4 has a frequency of 336 Hz what is the frequency of D_5?
    Answer: Calculate the interval G_4 - D_5. To find the frequency of D_5, an octave above D_4 we take the frequency of D_4 and multiply it by 2: $(9/8) \times 2 = (9/4)$;
    The interval G_4 - D_5 is: $D_5 = 396 \times \frac{3}{2} = 594\,Hz$

20.5 WHOLE AND HALF TONE INTERVALS

You will certainly be curious to know whether climbing up the just scale is like climbing up a ladder in which the rungs are equally spaced. Are all the intervals in this scale the same?

Easy to figure out if you remember that an interval is the ratio (rather than the difference) between two frequencies. So you can simply take a calculator and find the ratio between adjacent numbers, always taking the higher number divided by the lower. Below, we copy the just frequency table and underneath show the intervals:

Name:	C	D	E	F	G	A	B	C
ratio to C:	1	9/8	5/4	4/3	3/2	5/3	15/8	2
interval		9/8	10/9	16/15	9/8	10/9	9/8	16/15
in decimals		1.125	1.11	1.067	1.125	1.111	1.125	1.067
% change		12.5	11.1	6.7	12.5	11.1	12.5	6.7
Whole tone **W**, half tone **H**		**W**	**W**	**H**	**W**	**W**	**W**	**H**

Table 20.1 Increments between adjacent tones of the major scale. W is called a whole tone interval, or simply "a tone", H is called a half-tone or semi-tone.

If you want to get the answer as a fraction, rather than decimals, you need to remember how to divide fractions: you multiply by the reciprocal. For instance from F to G, the interval is

$$\frac{G}{F} = \frac{3/2}{4/3} = \frac{3}{2} \times \frac{3}{4} = \frac{9}{8}$$

We note two interesting results: the intervals are not at all the same, rather they come as two kinds: big intervals (called **whole-tone intervals**) and small intervals (called **half-tone or semitone intervals**). On the last line we label them as **W** and **H** (for whole and half, respectively). In Table 20.1 we also listed the frequency change from one tone to the next in percent change. For the whole-tone intervals the changes are 12.5% and 11.1%, while for the half-tone interval the change is 6.7%, nearly half of the larger intervals. So we may call the whole- and half-tones 1 and 1/2. Any sequence of tones that has this particular pattern of intervals:

Major scale intervals:

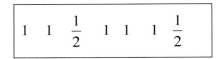

is called a *major scale*. Depending on the tuning of the instrument, the frequencies may differ a little bit from the just tuning, what is important here is that there are two kinds of intervals, large and small. Scales using two kinds of intervals, large and small, are called *diatonic scales*.

When you read in the concert program that a piece is in F-major, the composition uses a scale with a key note of F (the starting point of the scale) and again uses the above sequence of whole and half-tone intervals, starting on F. Major scales are widely used in western cultures, but less so in African or Asian music.

This is remarkable! If you compare the above sequence of whole and half-tones to a piano keyboard, you now suddenly realize why the black keys are distributed in the familiar pattern: the groups of two black keys and three black keys, separated by a gap, correspond exactly to the groups of two whole tones and three whole tones, separated by a half-tone gap. The black keys are where there is a large interval, **W**: the black keys provide a half-step where there is a big interval. We need no black key where there already is a half-tone interval because there already is a small interval at this place.

It is important to remember the sequence of whole and half tone intervals, but we now see that the sequence is easy if you picture the piano keyboard. *Where the black key is missing, two adjacent white keys are already at a half-tone apart.*

The musicians who developed the scale centuries ago may not have thought about it this way and presumably were guided by musical taste rather than numbers, but it is certainly not by chance that our tuning using triads quite naturally led to two kinds of intervals, half-tone and whole-tone, distributed in this particular way.

20.6 NAMES OF INTERVALS

Why is the interval from C to G called a "fifth"? Well, if you count all keys when you ascend the scale from C to G on a piano keyboard, indeed you find five, provided you include in your count the first key C as well. But it is somewhat strange to say that G is "a fifth above C", because it really is the <u>fourth</u> tone <u>above</u> C.

Having learned this quirk of musical nomenclature, it is now easy to deduce the names of other intervals and to find the frequency ratio for various intervals, if the keyboard is tuned according to just tuning. In Table 20.2 we list the names of intervals and their frequency ratio. We also count how many "semi-tones" the interval contains, where we count whole tones as 2 half tones (also called semi-tones). *The content of this table is not to be memorized. Rather, look at the piano keyboard and simply count the semitones!*

Remember, when we constructed the scale, we started with a "major triad", C – E – G. The interval C-E is a third, so is the interval E-G. But if you look at the keyboard or at the last row of Table 20.1, you may note that from C to E we go up 4 semitones (2 whole tones), but for E-G only three semitones. A third consisting of 4 semitones is called a *major third*, a third consisting of 3 semitones is called a *minor third*. The same idea is used for other intervals: C-D (1 whole tone = 2 semitones) is a major second; E-F (one semitone interval) is a minor second.

name of interval	number of semitones	examples (blue if just)	frequency ratio for just intervals
minor third	3	E – G, D – F	6/5
major third	4	C – E, F – A	5/4
fourth	5	C – F, D – G	4/3
fifth	7	C – G, D – A	3/2
sixth	9	C – A, D – B	5/3
octave	12	C – C, D – D	2

Table 20.2. Names of some intervals, number of semitones and just frequency ratios.

Table 20.2 shows that the just scale contains many intervals (ratios) that are consonant – simple number ratios. We calculate the ratio for instance for the minor third by looking up the frequencies for G and E in the just frequency table and dividing the higher frequency G = 3/2 by the lower one E = 5/4.

The adjective "just" is added to the name of a frequency interval if the interval has an EXACT simple number ratio such as 3/2. It is important to note that in "just" tuning not all intervals are "just"! Take the interval from D to A. It is a fifth, and consists of 7 semitones. But is it "just"? If we calculate the ratio of A = 5/3 to D = 9/8 we find - surprise! – 40/27, and not 3/2. More about this later, but remember that in just tuning, only those intervals that have nice simple number ratios and thus perfect harmony are called "just". For octaves one does not distinguish between just and non-just because octaves are assumed to have a perfect 2:1 ratio no matter what tuning is used.

In the third column of Table 20.2, we marked several examples of intervals that are not "just" – check it out yourself by using the numbers from the just frequency table.

Example. A 50-cm long string of a viola is tuned to G_4. How far from the neck must one place the finger on the fingerboard to play C_5? What is the name of the interval? (The "neck" is the far end of the fingerboard, where the tuning pegs are located).
Answer: Counting keys on the piano keyboard, the interval is as fourth. Check to see if it is a "just" fourth: the ratio is 2 divided by 3/2 = 2 × (2/3) = 4/3. It is a just fourth (Table 20.2). The new length of the string must be shorter than 50 cm in the 3/4 ratio. The new vibrating length is 50 × (3/4). The distance from the neck is what remains, 50 × (1/4) = 12.5 cm.

Diminished and Augmented Intervals

For completeness let us mention another problem of naming intervals. In Table 20.2 we said that a fourth has 5 semitones. If you try this out on the picture of the piano keyboard by placing fingers on C and F and moving the pair of fingers up the scale you find that indeed the fourth always spans one half-tone interval – with one exception! From F to B is indeed a fourth but an exceptional one, having three whole tones of 6 semitones instead of five. It is called the augmented fourth.
No such exception arises for the fifth, as you can easily convince yourself. But for the sixth you note that usually it spans one semitone and 4 whole tones, all together 9 semitones as shown in Table 20.2. There is one exception: from E to C is indeed a sixth, but this one is the exception, because it spans two semitones, ending up with 8 instead of 9. It is thus a smaller interval than the normal sixth and is called the *diminished sixth*.

20.7 TRANSPOSING - WHY BLACK KEYS?

If you look at the intervals of the "Frère Jacques" song at the beginning of this chapter, you will find that three tones correspond to **C–D–E** of the C-major scale, namely frequencies (with respect to C = 1) of 1, 9/8 and 5/4. Now if the pianist plays the song in C-major, using the white keys of the piano, the singer may find it a bit too high in pitch for his voice. He may ask the pianist to play her part "in a lower key" - starting a tone lower on B. The song needs to be "transposed" to a lower key.

When "transposing" we need to keep the same musical intervals, which means the same number of semitones between the various tones. Let us attempt to transpose a C-major scale down to B–major. The sequence of whole and half-tones: 1, 1, 1/2, 1, 1, 1, 1/2 must be maintained but we need to begin on B_3.

If we use the white keys of the piano, we are in trouble, because playing B-C-D instead of C–D–E is not what we want – it sounds not just lower but all wrong! C-D-E was two whole tone intervals in sequence, but B-C-D starts out with a semitone instead. The solution is to make a new tone, one whole tone above B. This tone will be a half tone above C – we use the black key to the right of C, tune it accordingly, and call it **C#** (called C-sharp). How do we tune C#? It must be a 9/8 ratio to B, because we want to maintain the original interval. What's the next tone after C#? D would not be right, because it is only a half tone above C#, so we use another black key, the one to the right of D and call it **D#**.

By now you should get the idea: to transpose a major scale, we keep the sequence of whole and semitones, 1, 1, 1/2, 1, 1, 1, 1/2, but start on a different tone and we use the black keys as needed to get the desired whole and half-tone steps. If you continue, you will find that the sequence of tones for the B-major scale is:

B-major scale: B C# D# E F# G# A# B.

Note that we did not need to calculate any frequencies. Instead we just sketch the piano keyboard and count number of whole tones and semitones. Music students have to memorize which black keys (also called "accidentals") to use for each scale. Here instead we learned to reason it out! A tone is called "sharp" or "flat" depending on whether it is the black key on the right or the left of the corresponding white key. So which of the names do you use? The rule is that you use the same letter only once – if the example above we would not replace D # by E b (E-flat).

20.8 PERFECTION SACRIFICED - THE TEMPERED SCALE

We talked about transposing and why there are black keys on the piano. Transposing from C-major to B-major required many black keys. Does the addition of the black keys take care of everything? There is a problem we swept under the rug thus far. We called the interval from _C_ to _D_ a whole tone, and from _D_ to _E_ a whole tone also. But these are not really quite the same. Look back to the chart of intervals in the Just Scale (Table 20.1). If I have tuned the piano to the just scale and want to transpose from C-major to D-major I am in trouble immediately, because the first interval of the major scale must be 9/8. But if I start playing the tune on _D_ the first interval is 10/9, which is a bit different. Maybe this is not a big problem because I don't play two adjacent tones at the same time anyway, so why worry about a little mistuning? But worse is to come: remember the beautiful, perfect, 3:2 ratio, the fifth. In just C-major tuning, _C_ to _G_ was a perfect 3:2 ratio, but if I start my tune on _D_ and play the tone a fifth above (_A_), the interval will be an ugly, out-of tune fifth – as we see if we work out the frequency ratio between _D_ and _A_. To figure it out, take the higher frequency (A) from the just frequencies in Table 20.1 divided by the lower frequency (D):

$$\frac{A}{D} = \frac{5/3}{9/8} = \frac{5}{3} \times \frac{8}{9} = \frac{40}{27} = 1.482$$

The fifth from A to D in Just tuning is far from a perfect fifth. Harmony depends on coincident overtones. Here, if we make A = 440 Hz, the frequency of D will be 440/1.482 = 296.9 Hz. Harmony requires that the higher partials coincide: here, the second harmonic if the A is 880 Hz, but the third harmonic of the D is 890 Hz. The 10-Hz difference produces a 10-Hz beat frequency, which we perceive not so much as rapid beats but as unpleasant roughness. For the same interval one octave or two octaves lower, all the frequencies would be half or a quarter as much, which means now the 10-Hz difference becomes 5 Hz or 2.5 Hz, producing very recognizable beats.

A very convenient way to picture this is to mark the frequencies of the scale on a **logarithmic** number line. To review the main point: on a log scale, the numbers are arranged in such a way that **equal ratios** (or equal factors) are represented by **equal distances**. This is perfect here, because equal ratios of frequencies means equal musical intervals, so that on a log scale **equal intervals** are shown as **equal distances**. This makes it easy to see what goes wrong in transposing. Transposing means to shift the entire scale up or down. In Figure 20.2 we show the just scale marked on a log scale (green marks) and underneath we show the same set of marks shifted to the right so that the first mark has moved from C to D. One notes that to play a scale starting on D, some of the green (C-major) frequencies are off, in particular the (green) E and A are too low in frequency, they are flat, while G and B are perfectly fine. This is confirmed by the calculation we did above for the interval from D to A. You might want to do the same for other intervals.

Help in visualizing scales:
- Equal musical intervals – equal frequency ratio
- On a "multiplicative" number line (= log scale) equal ratios are equidistant
- Advantage: in graphs below equal intervals have the same length

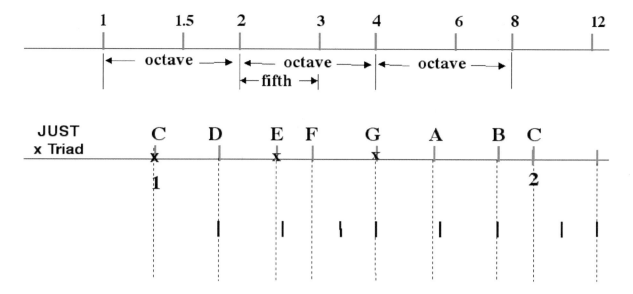

Figure 20.2. A logarithmic scale is used to explain the problem that arises in transposition when Just tuning is used. On the line labeled "Just" the frequencies of the just scale are marked (green marks). If we shift the scale up one tone (transposing from C to D) one obtains the short black marks below, as you can check yourself by copying the green scale and comparing to the black scale – they are the same but shifted. The short black marks indicate the frequencies we would need for a just scale in D. We see that G and B are tuned properly, but E and A would have to be re-tuned. The black marker between F and G and between C and D shows to what frequency the black keys F# and C# need to be tuned.

We conclude from this that no matter how we tune the two black keys, we still haveto retune the white keys when changing from C-major to D-major. Now we have two choices: either we stay with C-major and keep out of trouble or we re-tune the keyboard every half hour when we want to play a different key! To stay with C-major limits the musical expression. Although we said that the tune "sounds the same": when it is transposed to D-major or E-major etc., a different key makes it sound different just because it is higher, brighter, more sparkling.
Until the 18th century, keyboard instruments were retuned before switching e.g. from C-major to D-major. The black keys of course one can tune prior to the performance, but the white keys needed to be re-tuned. Depending on what key one wants to use, in some cases only a few tones are off and need adjustment. In Bach's time, music was played in very few keys and the keys were chosen so that between pieces only a few notes of the keyboard (harpsichord) had to be adjusted. But still, a great nuisance. Moreover, church music that included pipe organs was the driving force behind much of the musical creativity of great composers, among them Frescobaldi and J. S. Bach. Re-tuning organ pipes was and is completely impractical.

Composers and musicians have experimented with alternative tunings to alleviate these problems, and some of them were rather successful at least for applications limited to specific keys, such as Bach's version of tempered tuning. It is too complicated to describe these tuning compromises here, so we proceed to describe the contemporary radial solution.

Modern tunings use a strikingly simple scale, that is, a sequence of tone frequencies. Remember, the troubles started because in Just tuning, the whole tones are not all alike, some have 9/8, others have 10/9 frequency ratios. And in addition the semitone was not really half of a whole tone. That's why we had difficulty transposing from C-major to D-major. The idea of the modern *tempered tuning* is to *divide the octave into 12 identical semitone intervals*. Each whole tone then is exactly two semitones, and all are alike. Why 12 intervals? Because we want to retain as much of the Just tuning as possible, and in Just tuning each octave had 5 whole-tone intervals and 2 semitone intervals, which combined give $5 \square 2 + 2 = 12$ increments.

What would the frequency ratio have to be, then, for a whole tone and for s semitone? Remember that same interval means same frequency ratio. Each time we go up a semitone, we multiply the previous frequency by some number x, using the same number x 12-times so that all intervals are the same. But after 12 of these steps we must have the octave of the first tone. Expressed mathematically:

$$x \cdot x \cdot x \ldots\ldots \cdot x = x^{12} = 2$$

So now we need to figure out what number multiplied by itself 12-times makes 2!
The answer is:

$$x = \sqrt[12]{2} = 1.0595....$$

This is the semitone ratio of the tempered scale.

Tempered semitone frequency ratio: $x = 1.0595.....$	**(20.1)**

In the tempered scale the frequency ratio of any two adjacent keys of the piano (black or white) is the same and equals $1.059 \ldots = \sqrt[12]{2}$. The **advantage** of the tempered scale is that there is no difficulty transposing because the intervals between adjacent (black or white) keys on the piano are all exactly identical.

This only says that there is no loss when transposing without any need to retune – but it does not say anything about how badly out of tune the scale became, due to the radical change to equal semitone steps!

To see what happened we calculate the tempered ratio for various important intervals and compare to the desirable just ratios, as reported in Table 20.3.
The tempered whole-tone ratio is $x^2 = 1.1225$. This number is inbetween the two whole-tone ratios of the just scale. For the other intervals, the tempered ratios are easily calculated by multiplying $\underline{x}$ by itself a number of times equal to the number of semitones in the interval (or use the x^y key on your calculator).

name of interval	number of semitones	tempered ratio	frequency ratio for just interval	deviation %
minor third	3	1.189	6/5 = 1.200	0.9% - bad
major third	4	1.260	5/4 = 1.250	0.8% - bad
Fourth	5	1.335	4/3 = 1.333	0.12% - good
Fifth	7	1.498	3/2 = 1.500	0.07% - very good
Sixth	9	1.682	5/3 = 1.666	0.9% - bad
Octave	12	2	2	

Table 20.3. Comparison of tempered and just intervals. A deviation of 0.1% is still harmonious while one that approaches 1% used to be considered dissonant. These are labeled "good" and "bad", respectively.

Table 20.3 shows that the tempered intervals are not far from the just intervals. And most importantly, the strength of the tempered tuning is that the tuning is equally good (or equally bad!) for all keys, that is, one can transpose at will, without any further loss in harmony and no need for re-tuning. The *weak point* of the tempered scale are the relatively large deviations for the *major and minor third*, which are off nearly one percent, or 1 Hz for every hundred Hz in frequency. This means that for tones near 200 Hz, the tempered thirds are off about 2 Hz – enough to cause significant imperfection in the harmony. These tempered thirds were for a long time considered unacceptably dissonant, particularly because baroque compositions loved to use parallel lines between instruments spaced in thirds, but modern audiences no longer reject them. On the contrary, we are now so used to the "bad" tempered thirds, that some time ago when a music critic listened to a classical concert in Just tuning (without being aware of it) he complained that the performance was out of tune.

In Table 20.3 note that the tempered fifth, the most important of all consonant intervals other than the octave, deviates by less than 0.1% from the just ratio. Also the fourth is very satisfactory. In modern music terminology, the tempered fifth is referred to as a "perfect" fifth, even though it is not mathematically perfect.

A table of frequencies for tempered tuning with A_4 = 440 Hz is given in Table 20.4 at the end of this Chapter. A scale that, as the tempered scale, assigns a definite frequency (or frequency ratio) to all 12 semitones (white and black keys) is called a *chromatic scale*. For comparison, the frequencies of the just C-major scale are also listed in the table. The Just scale is not a chromatic scale, because it does not assign a unique frequency to the black keys – they are tuned differently as needed when transposing.

String orchestras or string quartets are not restricted to particular intervals. Books sometimes claim that good professional string quartets play the just scale for perfect harmony. Frequency measurements on first-rate professional string quartets show that they deviate from the just intonation and in fact play somewhere between the just and the tempered tuning.

Many musicians and music theorists are still searching for better ways to tune fixed-frequency instruments like the pipe organ. Not everyone is resigned that the modern tempered tuning is the best solution. Particularly, lovers of baroque music often use tunings other than the tempered tuning, and some musicians have made a name for themselves for their performances and recordings using historic baroque tunings.

Example: A guitar string is 60-cm long. The frets on the guitar are placed according to tempered tuning with a semitone from one fret to the next. What must be the distance from the neck (end of guitar) to the first fret?
Answer: the string must be shortened in the ratio 1:1.0595 (1 semitone ratio). The shortened string must be 60 cm/1.0595 long = 56.6 cm. This is 60-56.6 = 3.4 cm shorter than the original string. The first fret must be 3.4 cm from the neck.

20.9 MAJOR AND MINOR SCALES
As explained in Section 20.5, if you remember the layout of the white and black keys of the piano keyboard, you can reconstruct the sequence of whole and half tones:

Intervals of a Major scale: $1 \quad 1 \quad \frac{1}{2} \quad 1 \quad 1 \quad 1 \quad \frac{1}{2}$

As you might imagine, other sequences are possible. The next common choice results in the minor scale:

Intervals of a Minor scale: $1 \quad \frac{1}{2} \quad 1 \quad 1 \quad \frac{1}{2} \quad 1 \quad 1$

This is the sequence that occurs if you play the white keys of the piano starting on _A_. Musicians usually modify this scale, because the whole tones at the end of the scale sound strange, but we need not go into this here. If you play a minor scale it sounds very different from the major scale, even though it is the same sequence of whole and semitone intervals. We just started the sequence at a different place. Compositions in minor keys tend to sound more subdued, somber, even sad.

You can of course create new scales by starting the sequence in different places, such as D or E – these scales were used in Greek music and early Christian church music (called "Gregorian modes"). These go under the names of Dorian, Lydian, Aeonian etc. modes. Note that when we use the term "modes" to denote a particular sequence of whole tones and semitones (such as major mode, minor mode, Greek modes) this denomination has absolutely nothing to do with sinusoidal motion of string and pipe "modes".

Some modern composers use no particular sequence of whole and semitones, rather they use all semitone intervals within the octave (that is, all twelve white <u>and</u> black keys within the octave are considered equal). This is referred to as _twelve-tone music_. Remember that the intervals of the just scale came up with large and small intervals, that is, with whole and half steps, in order to produce harmonious intervals. In twelve-tone music, consonance (harmony) plays no important role. Some people like it nevertheless.

20.10 THE NATURAL SCALE

A sequence of frequencies that plays a special role in music is the so-called *natural scale*, also called the *harmonic series*. It is the scale that is played on a "natural horn" that is, a pipe without valves or tone holes (finger holes), playing only **multiples of the fundamental** frequency. Below we simply label the frequencies 1, 2, 3 etc. with the understanding that to obtain the frequencies in Hz, we multiply all of them by the fundamental frequency in Hz.

Natural scale	1	2	3	4	5	6	7	8	9	10	11	12
	C	C	G	C	E	G	X	C	?	10	X	G

Now we want to know how this scale relates to the just scale, which is also made up of whole numbers. You may recognize right away that the major triad (like C – E – G) is part of the natural scale. If the "natural instrument" is tuned to C, we can label the tones of the natural scale as shown: the first tone is C, so are the tones one, two and three octaves above (2, 4, 8). The 6th tone of the natural scale is a fifth (3/2 ratio) above the 4th tone (C), thus it is G. If the 6th tone is a G, so are all other tones an octave below or above (3, 12). Some natural tones don't match frequencies of the just scale (marked X). Bugle tunes use tones of the natural scale. What the tones are if the bugle is tuned to G rather than C you will be able to figure out by extending the frequency table for the just scale to other octaves. The 9th tone of the natural scale has a question mark because it does not match any tone of the just scale.

	Just		Tempered	
	Ratio	Frequency	Ratio	Frequency
C_4	1	264	1	261.6
$C^\#_4 D^b_4$			1.059	277.2
D_4	9/8 = 1.125	297	1.122	277.2
$D^\#_4 E^b_4$			1.189	311.1
E_4	5/4 = 1.25	330	1.260	329.6
F_4	4/3 = 1.333...	352	1.335	349.2
$F^\#_4 G^b_4$			1.414	370.0
G_4	3/2 = 1.5	396	1.498	392.0
$G^\#_4 A^b_4$			1.587	415.3
A_4	5/3 = 1.666...	**440**	1.682	**440**
$A^\#_4 B^b_4$			1.782	466.2
B_4	15/8 = 1.875	495	1.88	493.9
C_5	2	528	2	523.2

Table 20.4. Just and Tempered Scale: the table lists frequency ratios (with respect to C_4) and frequencies for the middle octave of the piano (assuming A_4 = 440 Hz for both scales).

Pablo Picasso
"Girl with a Mandolin" (1910)

Chapter 21

MUSICAL INSTRUMENTS

21.1 STRUCTURE OF MUSICAL INSTRUMENTS

Musical instruments are intended to produce sounds of a particular pitch and amplitude. In most cases, these sounds are more or less periodic over many oscillations and usually involve a complex wave shape that contains a number of partials, often 20 or more.

The physical structure of an instrument can be represented by the diagram in Figure 21.1. At the center of the instrument is an oscillating system (here called *resonator*), such as the string of a violin, the air column of a trumpet, or the drumhead of a kettle drum. To set the oscillator into motion, we need *an excitation device*, such as the bow of the violin, the lips of the trumpet player, or the piano hammer hitting the string. The excitation device in turn requires *a source of energy*, for instance the bowing arm of the string player, the lung of the flutist, the compressor that blows air into the organ pipe. Finally, the energy supplied by the energy source to the oscillator is given off as a sound wave, which requires some sort of *a radiator*.

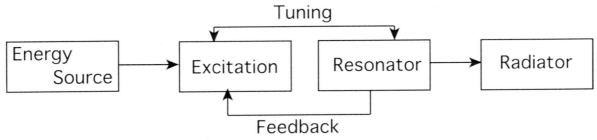

Figure 21.1. Schematic of the physical structure of a musical instrument

P.U.P.A. Gilbert and Willy Haeberli, Physics in the Arts, Part 2: Sound

For some instruments, the radiator is clearly visible to the eye: the sound board of a piano, the body of the guitar or the violin. The string itself is a very poor radiator because it is much thinner than the wavelength of the sound in air being radiated. Many brass instruments and woodwind instruments have a flared end (a "bell" or a "horn") to improve sound radiation. The sound radiation is usually stronger in particular directions, for the violoncello roughly perpendicular to the body of the instrument, for brass instruments primarily along the axis of the bell. Composers are aware of these technical details. For instance Bruckner specified in a loud trumpet passage that the trumpet is to be played "with raised bell". The flute does not have a bell, rather the radiator for the flute is the air escaping from the open ends and the finger holes, none of which are very efficient in radiating sound. Indeed, the flute in the orchestra struggles to be heard unless the composer gives it a short solo part now and then.

21.2 EXCITATION MECHANISM

For percussion instruments or plucked strings, the excitation is relatively simple: the oscillator is hit once and the excitation consists of a momentary impact. The excitation of a **steady tone** is more difficult: the oscillation sometimes must last for several seconds, which means that the resonator will make many hundred oscillations during a given tone. The musician has therefore a seemingly impossible task: he must control the excitation so that it is at resonance with the oscillator. To put it in simple words, suppose the trumpet plays an 880 Hz A_5 for three seconds. The air column in the trumpet during this time makes about 2600 oscillations. This then requires that the air column (the resonator) be pushed by the musician 2600 times, but with air puffs that are in step (in resonance) with the oscillator. By what magic does the musician accomplish this task?

The answer is that the trumpet player presses his lips against the mouth piece and blows to make a buzzing vibration of the lips. You can buzz the lips yourself even without a mouth piece by pressing the lips together fairly hard and blowing. When the musician first starts blowing into the trumpet, the pressure variation from the buzzing lips travels along the trumpet and is reflected again and again. It is the pressure wave from the end of the trumpet that locks the lip frequency to the pipe frequency, as long as the musician learned to buzz somewhere *near* the oscillator frequency. In effect, the excitation (buzzing lips) is the driving force on the resonator, but the resonator in turn affects the driving force (the lips) to keep it at the resonance frequency. This process is described as *feedback between oscillator and excitation*, Somewhat the same happens in organ pipes or a soprano recorder, in which air flowing through a slit is directed towards a sharp edge (Figure 21.2). The airflow is fast enough for the flow to make turbulence ("eddies" or vortices, the same you see in a fast flowing river behind a tree branch sticking out of the water). The oscillation of the turbulent air - the so called "edge tone" - sets in motion the air in the pipe. But the reflected wave from the end of the pipe in turn deflects the air stream, so that by feedback the oscillations of the air stream are locked- to one of the modes of the pipe. Here again, the frequency of the air oscillations by themselves must be close to one of the pipe modes. The recorder player can excite the second mode of the pipe by blowing harder and by starting the air stream very suddenly, because now the air stream oscillations tend to become faster and can be close to the second mode instead of the fundamental. The player "overblows" to the octave.

What we just described applies in a similar way to the flute, where the slit is formed by the player's lips and the edge is the sharp edge of the mouth hole. Almost everyone has at one time blown across the opening of an empty bottle to make it sound – again you create an edge tone that is close to the mode frequency of the bottle itself.

Figure 21.2. Air flows through a narrow slot rapidly enough to make turbulent eddies in alternation up and down. The periodic pressure variations set the air in the pipe into oscillation. The pressure wave reflected from the end of the instrument in turn affects the eddies (feedback).

Reed instruments, including clarinet, saxophone, oboe, and bassoon, have one or two strips of tapered, flexible pieces of cane or plastic (reeds) in the mouthpiece. Again these form a narrow opening through which the player's air stream passes. The reeds vibrate against one another (double reed instruments) or against the fixed mouthpiece (single reed instruments), interrupting the air stream. Again, the reed frequency is locked-in to one of the modes of the instrument. The player can choose which mode she wants to excite, by changing the lip pressure on the reed and by changing the speed of air flow. In this way she can, with practice, overblow the instrument to make the pitch jump from one mode to the other. Without the ability to overblow, the range of pitch available to be players of wind instruments would be far too limited.

For bowed strings, the musician pulls the bow across the string and sets the string into motion. Again, the driving force on the oscillator (the string) has to be in tune with the natural frequency of the string. How does the player wiggle the bow exactly 440-times every second when she bows the A-string? She doesn't! Again feedback comes to the help. The string player rubs rosin on the bow to make it slightly sticky. When she pulls the bow across the string, the string is pulled aside (Figure 21.3) until the restoring force is too large and the string loosens from the bow and snaps back. This pulse travels down the string and is reflected after time T, the roundtrip travel time on the string. The reflected pulse, when it arrives at the bow, helps to shake the string loose from the bow and the string snaps back at the right moment – so again the mode frequency of the oscillating system (string) feeds back in the excitation (bow) to keep it in tune. Again, the musician must learn to bow with the right pressure and speed of the bow to be near the desired frequency, otherwise you get the awful scratchy sound of a beginner's attempt to bow a string.

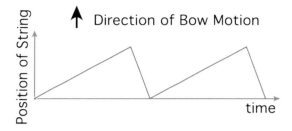

Figure 21.3. Schematic representation of a bowed string. The bow is made slightly sticky with rosin, so that the string gets pulled aside when the bow is pulled across the string. At some point in time, the string snaps back and the process repeats. The pulse reflected from the end of the string helps to shake the string loose at the right moment.

21.3 PLAYING A TUNE

To play a tune, the musician must be able to change the frequency of the sound. This requires changing the frequency of the oscillator to resonate at a sequence of frequencies, which correspond to the musical scale. This is a large and interesting topic, on which we make only a few brief comments..

1. **Strings:** strings are tuned to the right frequency by changing the tension with the tuning pegs. Once tuned, the frequency is changed by pressing the string against the finger board. This shortens the vibrating length of the string and increases the frequency in inverse proportion to the length of the string. All guitars have frets on their fingerboards. *The frets are placed according to tempered half-tone intervals.* Consequently, as you go up the fingerboard, frets get closer and closer, because for each successive half-tone the string needs to be shortend by the same fraction of the remaining length. When playing the guitar the finger is not placed *on* the fret, but between frets, so that the position of the fret accurately determines the length of the vibrating part of the string. Frets make it easy to play in tune (provided the strings were initially properly tuned with the tuning pegs), but have the disadvantage that they are not flexible. The frequencies are therefore fixed to tempered tuning and one cannot produce slides or a vibrato.

   In keyboard instruments, and also the harp, each tone has it's own string, properly tuned to the "right" frequency. One interesting aspect of piano tuning is that the octaves are not precise 2:1 ratios! This is related to string stiffness (Chapter 8.9) which increases the frequencies of the higher modes of a string a little bit. When you play two tones an octave apart on the piano, such as A_4 and A_5, you certainly want the second partial of A_4 to be in perfect tune with the fundamental of A_5. But because of string stiffness, the second partial of A_4 is a little more than twice the fundamental. Thus the A_5, tuned to the second partial of A_4, will be a little sharp, a little more then twice the A_4 fundamental – it will be a "stretched octave". For one octave, the effect is small, but by the time you keep doing this, one octave after the other, the errors accumulate. If one compares the lowest tone of a grand piano to the highest one about 7 octaves above, there is nearly a half-tone discrepancy compared to the mathematical 2^7 ratio. One can of course now synthesize sounds at will with computers – one can even play in just tuning - but it appears that the slight imperfections of real instruments are preferred by musicians and listeners.

2. **Woodwinds:** the lowest tone of a woodwind, such as a flute, oboe, bassoon, or clarinet, is played by covering all the tone holes. To play the tones of the musical scale, the player opens holes along the body of the instrument in succession starting from the bottom to shorten the air column and consequently raise the pitch. Proper placement of the holes is not as simple as placing frets on the guitar, because the pressure node in a wind instrument is not exactly at the first open hole. The holes simply cannot be made large enough to assure a pressure node. This is seen clearly if blowing a flute or a recorder you close a hole at the bottom end of the instrument, and keep an intermediate hole open. If the open hole had the effect of making a pressure node, with air escaping freely as if the pipe were cut off at that point, then closing a finger hole at the far end of the instrument should not change the pitch at all – but it does. Therefore placement of finger holes or tone holes and design of fingering patterns to keep the instrument in tune over three octaves is a challenge. All modern woodwinds use tone holes too large to be covered with a finger. Mechanically linked, padded covers over the tone holes ("keys") allow large tone holes and free choice of location of the tone holes. This, coupled with the ability of the player to overblow the instrument to higher modes allows about three octaves in half-tone intervals and pretty accurate tuning. It is known that Mozart only reluctantly composed for the flute on commission when he needed the money, because flutes of his time could not be played accurately in tune. If the modern flute (designed by Böhm based on physics calculations) had been available to Mozart, we probably would enjoy many more of his flute compositions.

3. **Brass Instruments:** Brass instruments include the trumpet, euphonium, trombone, French horn, etc. The air column in these instruments is never just a straight pipe, rather brass tubing is bent into beautiful curves because otherwise the instrument would be too large to handle. The same can be said for woodwinds playing low frequency tones, such as the bassoon, which is 10-ft long. The sound wave thus travels along a curved trajectory. As long as the bends are smooth and have no abrupt corners that would reflect waves, a curved pipe works just as well as a straight one, but is a lot easier to handle and to play.

Brass instruments change the frequency of the resonator by adding lengths of tubing. The slide used in the trombone is a simple and very transparent method to change the length of the pipe. Again the player needs to overblow to higher modes. Without overblowing the change in length of the air column needed to play three octaves would be impossible to achieve within the physical limits of a person's arm length.

Trumpets have three valves to change the length of the pipe. Other brass instruments sometimes have four or even five valves. When the trumpet player depresses a valve, a short length of tubing is added to the trumpet. To lower the pitch a semitone (ratio 1.059), we need to multiply the length of the trumpet by 1.059, that is, we add to the trumpet a little loop whose length is 5.9% of the original length. A second valve lowers the pitch by two semitones and a third valve by three. So all three valves depressed should lower the pitch by six semitones. But if you do the math you find that this is not quite true, because with increasing length of the air column, the amount of per semitone to be added gets larger. For this reason the trumpet is provided with a small slide operated by the thumb, which permits the musician to correct the pitch.

What comes after six semitones? The answer is seven, of course, but seven semitones is a fifth, and the trumpet player can produce a fifth by overblowing from the second mode ($2f$) to the third mode ($3f$).

Brass instruments are cylindrical pipes. They are simply tubes of a constant diameter to which a bell and a mouthpiece are added. The player makes a pressure anti-node at the mouthpiece. A pressure anti-node at one end is characteristic of a closed pipe, so we should *expect mode frequencies that are odd multiples* of the fundamental. But the higher modes of brass instruments are in fact $2f$, $3f$, $4f$, *as if they were open pipes*. The reason is that the higher modes of these closed pipes are shifted intentionally by proper design of the bell and the mouth piece to correspond to $2f$, $3f$, $4f$ etc. The frequency f is not actually a resonance mode of the instrument and is referred to as a "fictitious" fundamental. The book by Benade, mentioned at the end of the chapter, is a great source to read more about this.

Example: The air column of the French Horn is 2.6-m long. A piece of tubing is added by depressing a valve of the instrument. How long must the added piece of tubing be to lower the pitch by a semitone?

Answer: the semitone ratio of frequencies is 1.0595. The new length must be 260 cm × 1.0595 = 275.5 cm. The additional length is 275.5 - 260 = 15.5 cm.

4. **Voice:** when you sing a steady tone, air passes through the narrow opening between your vocal folds, which sets them into oscillations similar to the oscillations of the reeds on a woodwind. The singer can control the frequency of these oscillations by muscles, which change the tension on the vocal folds (also called "vocal cords"). But the air pressure variations are far from sinusoidal, rather the opening and closing of the vocal folds makes a sequence of sharp, periodic puffs of air with a large number of Fourier components in the sound spectrum. Air in the throat and in the mouth ("oral cavity") is set into oscillation by these puffs of air. Once more this is an oscillator set in motion by a driving force. As far as resonators go, the throat and oral cavity have a large amount of damping and thus a wide resonance curve. The only components of the vocal folds sound spectrum that are enforced (are strong), are the ones that resonate in the throat or in the oral cavity. In Chapter 19, you see a spectrum of Willy's voice. The vocal folds produce 220 puffs of air every second, so that the partials are all multiples of 220 Hz. The sound spectrum shown in Figure 19.7 shows that the resonance in the oral cavity reinforces some frequency regions of Willy's voice. *These are called the formants of the particular vowel. Singing a **different vowel** of the **same frequency** would have the **same set of partials** (multiples of 220 Hz) but **different formants**.*

5. **Percussion Instruments:** Drums, bells, xylophones etc. are made to oscillate when struck or hit by hands or by sticks. Large drums are the most powerful orchestral instruments – a single drum can compete with an entire orchestra in sound power output. For drums, the oscillations are heavily damped and thus decay rather rapidly. A special property of percussion instruments is that the *higher partials are non-harmonic*. Since the tone does not last very long, non-harmonic partials are not

perceived as unpleasant. Some Indian drums (the Tabla) use drumheads loaded by adding sticky material near the center of the drum, which permits shifting of the higher partials to be nearly harmonic. For bells and cymbals, the tone may last for a second or more, but the higher partials decay quickly so that very soon only the lowest partials are heard. These are tuned to be nearly harmonic.

Drums and bells have modes like strings: motions in which all parts of the instrument move at the same frequency, and therefore undergo simple harmonic oscillation. As an example, modes of a drum head are shown in Figure 21.4, in which the dashed lines show the positions of node (nodal lines). Along one of the nodal lines, the drum head does not move, as one can show by sprinkling sand or sesame seeds on the drum. On either side of a nodal line, the drum head oscillates in opposite phase, just like on a string the motion is opposite to the left and the right of a node. To excite a drum head to a single mode one can use a loudspeaker placed near the drum head in a place where the desired mode has a antinode. The formula in Figure 21.4 shows similarity with the string formula. Here the quantity F is the pull per unit length of the circumference, σ is the mass per unit area of the drum head, and D it's diameter.

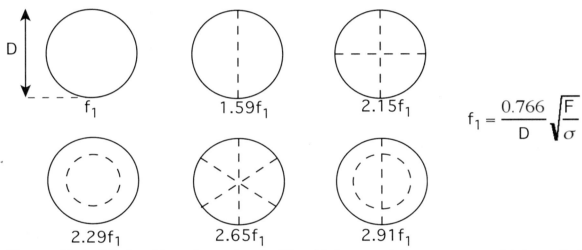

$$f_1 = \frac{0.766}{D}\sqrt{\frac{F}{\sigma}}$$

Figure 21.4. Modes if a drum head. The higher partials are not multiples of the fundamental, that is, the partials are not harmonics". The frequency formula is explained in the text.

H. Benade, "Horns, Strings and Harmony" (Dover Publications).
D. A. Hall "Musical Acoustics" Third Edition, Brooks/Cole publishers (2002)
J. G. Backus "The Acoustical Foundations of Music" 2nd Edition, W.W. Norton (1977)T.
D. Rossing "The Science of Sound", 2nd Edition, Addison Wesley (1990)